traditional
societies
and
technological
change

# traditional
# societies
# and
# technological
# change
SECOND EDITION

**GEORGE M. FOSTER**
University of California, Berkeley

Harper & Row, Publishers
New York   Evanston   San Francisco   London

# contents

Preface to the Second Edition     vii
Introduction     1
1. The Cultural Context of Technological Development     9
2. The Rural Community: The Traditional World     25
3. The Rural Community: The Contemporary World     42
4. The Dynamics of Change: Culture, Society,
    Psychology, and Economics     76
5. Cultural Barriers to Change     82
6. Social Barriers to Change     105
7. Psychological Barriers to Change     130
8. Stimulants to Change     148
9. Bureaucracies and Technicians     175
10. The Anthropologist at Work: The Conceptual
    Context     197
11. The Anthropologist at Work: Stages of Analysis     215
12. Technical Aid and Behavioral Science: Some
    Problems of Teamwork     233
13. Ethics in Planned Change     246
Works Cited     260
Index     275

# preface
# to the
# second
# edition

Since the publication in 1962 of *Traditional Cultures: and the Impact of Technological Change*, the world has changed greatly, and so have our understanding of and our approach to the problems of aiding developing nations. Numerous former colonies have acquired their independence, and along with it the responsibility of providing for the welfare of their people. Notable industrial development has occurred in some nations, such as Mexico and India, and almost all countries have made significant progress in food production, health, and educational services. Particularly encouraging is the Green Revolution, the long-awaited breakthrough in agricultural production, which promises, at least for a time, increased nutritional levels for the world's hungry. Thanks to the founding of new national universities and vast improvements in older ones, developing countries are beginning to have sizable pools of trained technical specialists. And these human resources have been augmented by continuing streams of young people who have studied in Europe and North America before beginning work in their own countries.

The most striking phenomenon within these countries has

been the rural exodus, the rush from village to cities. Although urbanization in developing nations began on a significant scale a generation or more ago, only during the last decade has the magnitude of the process, and its implications for orderly national development, become fully apparent. The urbanization process in turn is largely a consequence of unbridled population explosion: Rural areas can no longer provide work and opportunity for growing numbers of people, who are forced to seek employment in cities or migrate beyond the boundaries of their countries to earn money in foreign factories and fields.

With respect to the methodology of technical assistance, we have learned a great deal. We have also been sobered. We now know that what at one time seemed relatively easy to stimulate—technological and economic development—is an enormously complex and often discouragingly slow process. The right mix of money and skilled specialists in health, agriculture, and education—valuable as far as it goes—is insufficient to solve the problems that face the world. Disillusioned by frequent lack of progress and by the apparent ingratitude of some of the nations we had hoped to help, Americans have become less willing to support foreign aid programs. Moreover, a great many Americans have been further disillusioned by the use of foreign aid in southeast Asia to support military activities. On the positive side, we have finally faced up to the implications of unrestrained population growth, and it is now permissible not only to discuss family planning, but even to finance research and action programs from public funds.

These, and other comparable developments, suggested the need for a major revision of *Traditional Cultures*. In this new edition, several chapters have been completely rewritten, others have been extensively changed, and a few have received only modest editing. Chapter One has been enlarged to include the concept of the premises that underlie cultural forms and individual behavior. Chapters Two and Three have been completely rewritten, the first to incorporate my model of *limited good* to explain traditional peasant behavior, the second to explore the nature and implications of both urbanization and the rising sense of nationalism found in all developing countries. Chapters Four through Eight have been thoroughly updated, but the basic pattern of describing change in terms of barriers and stimulants has been retained. Chapter Nine retains unchanged the section on *culture shock*, but almost all of the remainder is new, reflecting my thinking about the importance of understanding and studying bureaucracies just as we study the recipient peoples. And the ego-gratification needs of technical specialists are discussed more fully than in the first edition. Chapters Ten and Eleven, which describe how the anthropologist works in technical assistance programs, and Chapter Twelve, dealing with problems of teamwork between administrators and anthropologists, have been brought up to date, but changes have been moderate. Chapter Thirteen, which focuses on the ethics of planned change, has been completely rewritten to reflect the growing concern among anthropologists about the ethical implications of their work.

The present book represents the fruit of more than a generation of research and applied consultative experiences, and the stimulation and aid of hundreds of friends from all parts of the world. I hope that those of them who may read this new edition will be reminded of our pleasant associations, and that they will feel I have indeed learned much from them.

George M. Foster

# traditional
## societies
## and
## technological
## change

# introduction

Until very recently, most of the world's peoples have lived in "traditional" societies. Many still do. Although traditional societies display significant social, cultural, and linguistic variations, in a broader sense they constitute a generic type, because they share common structural and cultural characteristics. By "structural" I refer to the historic pattern of concentration of economic and political power in the hands of small elite groups who exploit their position to maintain the status quo and to ensure that they and their descendents continue to enjoy the good things of life, at the expense of the masses, both rural and urban. At the cultural level, traditional societies are marked by a variety of traits. Technologies are simple, even for the elite, and scientific knowledge is limited. Production is achieved through human and animal, rather than mechanical, power. Most agriculture is unproductive: Human or animal manure may be used, but chemical fertilizers and pesticides are unknown. Consequently, yields are low and subject to the vagaries of weather. Environmental sanitation is poor, medical services are primitive, chronic illness prevails, life spans are short, and both birth and death rates are high. Illiteracy is the rule, and access to new knowledge is restricted. For the masses, these conditions mean poverty and uncertainty and limited personal freedom. In most traditional societies the masses have had few, if any, options.

Throughout history people in traditional societies have accepted these conditions because they knew nothing else. Revolutions, while they may loom large in historical perspective, have been the exception rather than the rule. Certainly the numbers of peasants who have personally been involved in revolutionary activities must be a tiny fraction of those peoples who have made up traditional societies. This historic resignation no longer holds, however. Although the elites may be quite happy with the status quo, the rural and urban masses of traditional societies are no longer willing to continue under the old rules of the game. Increasingly they demand release from the exploitation and grinding poverty that have been their lot for centuries, and they expect a greater return for their labor. They desire better health and medical services, schooling for their children, new consumer goods, and the right to move to cities in search of a better life. They want, in a nutshell, to exchange traditional society for contemporary society.

Although social scientists argue the details of how traditional societies are transformed, they agree on the major requirements: Feudal power structures that concentrate land, wealth, and power in the hands of an hereditary elite must go; new political and social forms are essential to more nearly equalize opportunity; governments must be more responsive to the needs of their peoples, more ready to act on the basis of the greatest good for the greatest numbers. But new power structures and political and economic institutions, alone, are not sufficient to the task. Distributing worn-out and nonproductive lands to peasants ignorant of the possibilities of scientific agriculture is only a modest favor. Equality of access to a static economic pie is a step forward from inequality of access, but it is insufficient by far to meet the demands of contemporary traditional peoples, and it is unnecessary as well. With the rational use of capital, of labor, of resources, and of knowledge, the pie can be expanded, if not indefinitely, far beyond previous sizes.

In *Traditional Societies and Technological Change* I am concerned with that aspect of this expansion of the pie that we usually call "technological development." I am interested in the processes whereby better machines, mechanical power, and new scientific skills and knowledge can be introduced into developing countries— made common property of the peoples of these countries—and thereby help make it possible for them to enjoy some of the advantages that those of us in more fortunate circumstances have taken for granted. I am concerned particularly with what have been called "human factors in technological change," or "social aspects of technological development," as they are manifest in today's changing traditional societies.

In the industrialized world—particularly northern Europe and the United States—technological development was largely the product of self-generating forces, the gradual accumulation of scientific and organizational knowledge, and the evolution of a new type of urban life. The process was relatively slow, which probably was a great advantage. Control of disease lagged well behind other forms of technical progress, so that an industrial base with abundant employment opportunities had been created prior to the appearance of a wide gap between birth rates and death rates. Unemployment and underemployment, while not unknown, were not a major drag; more often, as in the United States, there were insufficient hands to man the new machines.

In today's developing countries the strategy of national development is quite distinct: Since there is no need to rediscover fundamental scientific laws nor to reinvent the techniques of applying them, the technological knowledge of the Western world can quickly be transferred to new countries. On one level the problems seem simple. The administrative, agricultural, health, and educational skills needed to satisfy the new needs of people in previously traditional societies can easily be taught, so that with help from wealthier nations every country should be able, in a generation or two, to develop human and other resources adequate to its needs. But on another level the problem has proved to be far more difficult. Spectacular success in transferring disease control techniques from industrial to developing countries has produced, in the form of population explosion, the greatest of all threats to new nations, and to the world as a whole. Burgeoning villages whose populations are no longer held steady by high death rates send forth their surplus members to seek employment in cities, where employment opportunities in nascent industries lag behind the demand for work. How can developing nations buy time to adjust their populations to the opportunities for productive employment? Can they, with all the technical assistance in the world, do so? We do not know. We do know that time is short, and that the entire world has a basic stake in a successful outcome.

Recognition of the need for, and the inevitability of, the transformation of traditional society has led, since the end of World War II, to a major national and international effort, technical assistance, whose goal is to speed up technological development in—you may choose your euphemism—"underdeveloped," "less developed," or "newly developing" countries. Initially we believed that the transmission of new technical skills to needy peoples was a simple process, that all that was necessary was to make available to them the scientific and practical answers that had worked well in industrialized countries. Now we have learned that the process is much more complex than we had thought, and that there are cultural, value, and ethical, as well as simple technical, dimensions to the process. Part of the problem is that the expression *technological development* is itself misleading; strictly speaking, there can be no such thing as technological development in isolation. Perhaps the use of the term *sociotechnological development* would clarify our thinking, for development is much more than the overt acceptance of material and technical improvements. It is a cultural, social, and psychological process as well. Associated with every technical and material change there is a corresponding change in the attitudes, the thoughts, the values, the beliefs, and the behavior of the people who are affected by the material change. These nonmaterial changes are more subtle. Often they are overlooked or their significance is underestimated. Yet the eventual effect of a material or social improvement is determined by the extent to which the other aspects of culture affected by it can alter their forms with a minimum of disruption. In newly developing countries, for example, the introduction of factory labor brings changes in family structure. If the workers and their families can accept these social patterns and reconcile their attitudes toward traditional family obligations with new conditions, industrialization need not be disruptive. But, as we shall see, such reconciliation is often difficult, and the process of development is accordingly slowed.

At present we know much less about the cultural, social, and psychological aspects of development than about the purely technological. A public health physician knows he can successfully plan the clinical aspects of a campaign to stamp out smallpox. The problems of the production of vaccine and its storage and transportation to the places where it is needed have long been solved. But the physician does not know how he can induce everyone in a village to come forward willingly to be vaccinated. Force has frequently been resorted to in such campaigns, against the will of terrified people who do not understand. An agricultural extension agent can analyze soil conditions and prescribe a hybrid seed and modified cultivation practices that will vastly increase production. But he cannot be sure that farmers will see the benefits he is sure lie in his improved methods; often he is puzzled by lack of interest in "obvious" advantages. It is a simple technical problem to teach an adult to read, but it is quite another matter to make the adult want to know how to read or to create an environment in which it is to his continuing advantage to do so.

We who, through education and opportunity, share the values of the complex civilizations of the West, tend to think of development (which we equate with progress) as manifest in science and technology: better automobiles, faster airplanes, finer buildings and more comfortable homes, hybrid seed and larger crops, and miracle drugs and better health. The ingenuity of scientists and technicians captures our imagination; it is hard to doubt that the innovations they produce are good, and it is easy to assume that these contributions to better living must appear equally desirable to people in every part of the world, once they are aware of them.

It sometimes comes as a surprise, then, to find that many people in technologically less advanced lands are reluctant or unable to accept change with the same ease we do. In spite of their desire for new opportunities and new consumer goods, they often do not realize, or are unwilling to accept, that the goals they strive for can be achieved only at the cost of major sacrifices of old values and customs. The wisdom of tradition still carries weight among many of them, and the cries of "new" and "better" may set people on guard rather than stimulate their desire to experiment. The urge for development and the willingness to change are not equally present in all peoples. What may seem desirable to one person may seem undesirable to another, and what may be good for one may be bad for a second. In all of us, motivations to change are opposed by resistances to change. The factors that determine these motivations and resistances are cultural, social, and psychological. They may be rooted in the value system that characterizes our culture; they may be associated with the nature of relationships among the members of our group or with problems of status and role; they may lie in faulty communication within a group or between members of different groups (as, for example, between technical advisers from one country and villagers of another); or they may be found in any of a vast number of other nontechnical contexts.

In recent years, the existence of human factors in technological development has been increasingly accepted. It has been generally recognized that technical experts who work in programs of international aid do better if they understand something about the cultural and social forms of the groups to which they are sent. Lip service, at least, is commonly paid to this axiom in briefing specialists for field-work. On this level the human problems generally are thought of as falling in two categories: (1) the need to acquaint the technician with the language and culture of the country to which he goes so that he will know the correct forms of social intercourse, so that he can play the role of a gentleman and avoid giving offense through ignorance; and (2) the need to know basic social and cultural forms so that the technician will understand why villagers may fear vac-cination, why they often resist new agricultural programs, why lit-eracy campaigns sometimes fail to attract them. With this informa-tion the technician can seek ways to overcome resistances.

This is good, and correct, as far as it goes. But a thorough un-derstanding of the cultural forms and values of the people of an-other nation is not in itself sufficient to eliminate human problems. In developmental programs representatives of two or more cultural systems come into contact. Sometimes the differences are major, as when an American physician works in India. Sometimes the differ-ences are less pronounced, as when a city-bred-and-educated Indian physician is charged with health operations in a rural area of his country. But whether the gulf separating the two worlds is full-cul-tural or subcultural, it is significant. In either case, the technician shares the cultural and social forms not only of the country from which he comes, but also of the professional group he represents. *His* basic attitudes and beliefs, and the implicit premises that under-lie his behavior, stem from his cultural conditioning in the same fashion that the behavior of the people he hopes to influence stems from their cultural conditioning. Whether he is a doctor, an agri-cultural extension agent, or an educator, the specialist is profoundly —and often unconsciously—influenced by the value system of his professional subculture. From his student days he has been indoc-trinated with the motivations, the goals, the ideas about the right ways to work and the ethics of his field, and he has learned a par-ticular role in the bureaucratic hierarchy through which his profes-sion is organized for action. The technician's uncritical acceptance of his professional point of view, as well as of the underlying assump-tions of his culture, can be as much a barrier to change as can the cultural forms of the target group.

There is still a third type of human problem in technological de-velopment: the ethical one. How far—if one may do it at all—is it right to go in deciding what is good for someone else? Do education and technological competence confer upon the technician wisdom to decide what other people should have? Does the state, through its professionally oriented technical services, have the right or obligation to make basic decisions that will profoundly change the lives of its citizens? Or—to take a real problem in developmental work—should villagers in newly developing countries remain untouched until they decide, by democratic processes, that they are ready for change? There is no easy answer to the ethical question, and certainly no answer that can claim to be scientific. But wide recognition of the ethical problems inherent in developmental work, and discussion of their implications, will help in making the necessary decisions in every specific program.

Thousands of trained American technical specialists, some well oriented and others less well oriented to the meaning of social, cultural, and psychological factors in technological change, have been sent overseas since the end of World War II. Their efforts have been augmented by those of equal numbers of Peace Corps Volunteers. For all of them able to learn there has been, in the words of Cousins, "the dramatic discovery of conspicuously different cultures" (Cousins 1961:30). This is always a heady experience, but it may also be an unsettling one. American technical specialists work with the citizens of many countries in the free world, sometimes in the offices of their technical bureaucracies, sometimes at grass roots levels in villages, on farms, and in the slums of large cities. Frequently they are powerful agents for international understanding and goodwill while they are simultaneously effective in transmitting their skills or services. This is most likely to be the case when they are sensitive to the implications of cultural differences and when they recognize that the basic values and the implicit assumptions of American life do not automatically provide answers to the dilemmas that face them. For all, there is culture shock, sometimes silent and insidious and sometimes breathtakingly brutal—as when a young American doctor in a Bombay hospital found himself, for the first time in his life, questioning his fundamental values:

> I just wasn't psychologically attuned to the problems I
> would have to face. Back in the United States a doctor
> never has to ask himself: 'Why try to keep this baby alive?' He
> concentrates all his knowledge and will power on the need
> to save a child and give him a chance for a normal life. But
> here, in India, maybe two or three hundred million people will
> never experience a single day free of hunger or sickness
> in their entire lives (Cousins 1961:30).

This book deals with the cultural, social, and psychological sides of technical assistance and technological development, particularly as it occurs in traditional rural communities that have been the targets of developmental programs such as those of the specialized agencies of the United Nations, the United States bilateral aid programs, and programs sponsored by private organizations like the Rockefeller and Ford foundations. It attempts to place the phenomenon of planned technological change in the broader perspective of the underlying processes of culture change, which occurs at all times in all parts of the world. Reflecting the academic bias of the author, the interpretation is narrowly anthropological and more broadly sociological. The nontechnical barriers to change found in all societies are discussed at length, as are the factors that stimulate change. Then the focus is shifted to the American technician, his professional culture, the relation of his training to the task at hand, and the experience of culture shock, which all of us undergo when first working abroad.

I believe that many of the difficulties that plague technical specialists in overseas assignments and prevent the most effective developmental programs can be ameliorated by an understanding and a better utilization of social science, and particularly of anthropological knowledge. Accordingly I devote the final chapters to a discussion of how the social scientist works in an applied setting and of the kinds of administrative relationships that make possible the best teamwork between program administrators and social scientists. Finally, I touch briefly upon some of the factors that seem to me to be involved in the ethics of developmental work.

This book is not intended as a handy pocket guide to successful technical aid work; it contains no formal lists of dos and don'ts. It is my hope, however, that the ideas and the evidence presented will contribute to a better understanding of the implications of planned technological change and the role of the personnel who participate in such programs, and that the practical hints given from time to time will prove useful.

# 1

# the cultural context of technological development

When we speak of *technological development* in newly developing areas, what exactly do we mean? It is clear that the recipients of public health aid, agricultural extension projects, and community-development programs are learning novel techniques, acquiring facility in the handling of new tools, and changing attitudes and customs that impinge on many areas of their lives. But technical specialists differ from social scientists in the frame of reference they use to interpret what is happening. The technical specialist, with the goals and values of his professional subculture, sees developmental work as the successful diffusion of scientific knowledge and of behavior based on scientific knowledge to areas where they previously existed in slight measure or not at all.

The social scientist, on the other hand, as a product of his academic subculture, looks upon technological development as a particular kind of change in the structure of society, in patterns of culture, and in individual behavior. He describes what happens as "planned" or directed" or "guided" change, to distinguish the process from the evolutionary or spontaneous or unplanned changes that constantly occur.[1] In order to understand how these domains of

[1]Revolutionary change, such as that of Kemal Ataturk's Turkey or of Communist Russia or China, is also planned change. Not only is it planned change,

change interrelate, I find it helpful to speak of "The Three Systems."

Men and their activities in social groups can be analyzed into three basic systems: first, a system of social relations, in which persons and groups are linked together by rights and duties, by expectations and obligations; second, a system of group conduct, which predisposes men to think and behave in normative ways according to their perception of circumstances; and, third, a system of individual cognition and behavior, which underlies the first two systems and which, rooted in biology and life experience, determines how the individual will react in a given situation. The three systems, the social, the cultural, and the psychological, are called, for short, *society, culture, and psychology.*[2]

Since the terms *society* and *culture* are often used interchangeably, it is important that we know what each one means. Fortunately we need not concern ourselves with the elaborate definitions that anthropologists and sociologists have coined for the systems they study. For our purposes we can define culture and society and illustrate their interrelationships sufficiently well by considering the people with whom we work in developmental programs. Technical specialists direct their efforts toward groups of people; they work in communities, often as easily recognizable as a village or small town. These communities are not simply crowds, agglomerations of people who chance to be physically close to one another. They are societies, organized groups of people who have learned to live and work together, interacting in the pursuit of common ends. In order to live and work together people require formal rules for defining their relationships one to another. We say that a society has a *structure,* or an *organization,* a patterned arrangement of relationships that social scientists discern in the behavior they observe, a system that is seen to define and regulate the mode of everyday intercourse among the members of the group. Social structure usually is studied by breaking down its parts into *institutions* such as the family, religion, political and economic systems, and the like, each of which harbors a great number of *statuses*—the positions people occupy vis-à-vis each

---

but it is very effective planned change, perhaps the most successful of all, if our measure is degree. Nevertheless, because of its very different ideological framework and very different techniques for modifying behavior (as compared with the kinds of change with which we are dealing in this book), I find it undesirable to try to combine both variants in a single treatment.

[2]In speaking of "Three Systems" I have paraphrased and added to Freedman's "Two Systems" introduction to his very stimulating "Health Education: How it Strikes an Anthropologist" (1956).

other, which in turn have a corresponding *role*, the behavior forms deemed appropriate to each status.

A particular society is a going concern—it functions and perpetuates itself—because its members, quite unconsciously, agree on the basic rules for living together. *Culture* is the shorthand term for these rules that guide the way of life of the members of a social group. More specifically, culture can be thought of as the common, learned way of life shared by the members of a society, consisting of the totality of tools, techniques, social institutions, attitudes, beliefs, motivations, and systems of value known to the group. Or, to put the distinction in a different way, *society* means people, and *culture* means the behavior of people. The terms are interdependent; and it is difficult to speak of one without relating it to the other. Social scientists often use the terms interchangeably, or they use the compound *sociocultural* to indicate that the phenomena dealt with partake of both society and culture. To illustrate, we can say that "in peasant society prestige is an important motivation in bringing about change," or "the prestige motivation is recognizable in peasant culture as an important factor in change," or, yet again, "among the sociocultural factors that explain change in peasant communities, the desire for prestige is high on the list." All three sentences convey essentially the same idea.

The terms *culture* and *society* have both specific and generic meanings. We speak of "a society," a particular, recognizable, delineable, finite body of people; it may be a primitive hunting band, an agricultural village, a market city, or a modern nation. And we speak of "a culture," which is, of course, the particular way of life of a specific society. We also use the terms, sometimes in the adjectival forms of "social" (or "societal") and "cultural," in a nonspecific, generic, panhuman sense. We can speak of social and cultural forms of mankind without reference to specific groups, and we can study the evolution of culture and the development of society in an equally generalized sense.

For the sake of simplicity let us say that technological development programs represent planned sociocultural change, recognizing the psychological dimension that always underlies change processes. As technical specialists we hope to help specific groups of people to change some of their behavior forms, to modify their sociocultural systems. First, of course, it is highly desirable that we have some understanding of ourselves as social, cultural, and psychological beings. This problem will be explored in Chapter 9. And, second, if we are helping to change a sociocultural system, it is wise to have a pretty good idea of what it is that is changing. So now we shall consider the basic characteristics of sociocultural systems—of societies and cultures—particularly with respect to those points that relate directly to the process of change.

1. *Sociocultural forms are learned.* The behavior patterns found in a specific sociocultural system are not genetically or biologically determined. Every normal infant has the potential to learn any culture and to work effectively as a member of any society. The particular sociocultural forms that an individual comes to share is a matter of chance, the accident of his place of birth. Through the process of socialization, or enculturation as Herskovits calls it, the child acquires the prevailing attitudes and beliefs, the forms of behavior appropriate to the social roles he fills, and the behavior patterns and values of the society into which he is born. Because culture is learned rather than transmitted biologically, it is sometimes called man's social heritage.

If culture stemmed from race, change could occur only as biological forms change; for practical purposes, directed progress would be impossible. But because of the innate potential of the human infant and the inherent plasticity of the human mind, man not only learns a culture and its associated social structure, but he can also forget or cast aside parts of a culture and adopt in their place new and often radically different behavior forms.

Although everyone learns the culture into which he is born, the human environment and life experiences of no two people are absolutely identical. The chance that determines one's society also determines one's statuses within that society, and these in turn affect the individual learning process. Consequently, although a culture or society produces similar products, these products are not absolutely uniform. No two people act and react in exactly the same way. Each person has a unique personality—"the organized aggregate of psychological processes and states pertaining to the individual," to quote Linton (1945:84)—and this personality, in relation to the society and culture that go with it, will determine the specific behavior of its owner. The fact of individual variation among the members of the same social group has important implications for planning change. Some people will be psychologically more disposed to try new things than will others. Depending on life experiences, there will be differences in the ease and ability with which villagers continue to learn and in their flexibility in casting off old forms of behavior that conflict with new forms. A technical expert quickly learns that his first practical task in a new setting is to identify the relatively few individuals in the community whose personalities predispose them to sympathetic consideration of his program.

2. *A sociocultural system is a logically integrated, functional, sense-making whole.* It is not an accidental collection of customs and habits, of roles and statuses, thrown together by chance. If the analogy is not carried too far, a sociocultural system may be compared to a biological organism, or to an ecological system, in that each of its parts is related in some way to all other parts. Each fulfills a definite function in relation to the others and is essential to the normal functioning of the system as a whole. Each part, in turn, draws upon all the other parts in some way for its own continued existence, and its growth and development are dependent upon corresponding growth and development in the system as a whole. To put the matter another way, each institution—religion, for example—reflects the dominant values of the total culture, and the beliefs and activities that constitute religion articulate with the system at a thousand points. Social, economic, and juridical phases of cultures and societies cannot be fully understood without an understanding of religious forms, which in turn are expressed through special speech patterns, social rites, mythology, music,and material culture.

To say that a sociocultural system is a logically integrated, functional entity does not imply that its parts interact in perfect harmony, without stress or strain. In an absolutely static community this might be true. But cultures change, and the parts of a system change at different speeds; consequently, perfect integration and perfect fit are impossible. Varying degrees of disorganization and apparent logical inconsistencies will be found in any specific situation. Every culture, therefore, represents something of a compromise, an attempt to strike a balance between the stresses and strains that are inevitable consequences of unequal rates of change and the forces that work toward the unattainable goal of perfect harmony. Preexisting relationships between cultural and social elements are constantly disturbed by normal processes of change, and, just as constantly, the affected areas strive to adjust themselves to new circumstances and to reach a new stable relationship that may endure until still other steps forward are taken.

It should be apparent that it is important for the practitioner or student of technological change to understand a sociocultural system as a functional, sense-making, logical whole. Significant change in any social institution or phase of a culture cannot occur without accommodation in those institutions or phases that impinge upon it, and the degree of possible change is limited by the extent to which these accommodations occur. Conversely, any change in one institution produces secondary and tertiary changes in others, of a nature and extent that cannot always be foreseen.

Historical changes among the Tanala, a hill tribe of western Madagascar, illustrate this truism. Until the early eighteenth century slash-and-burn dry rice cultivation forced villages to move every few years to find fertile land. Under these conditions individual ownership of land was not necessary, and no marked inequalities in wealth developed. Joint or extended families owned growing crops and worked together as social and economic units, but land reverted to communal ownership when a village moved on. When wet rice appeared in small plots of wetland in valley bottoms, single-family cultivation was adequate to the task, and the joint family began to break down. Then followed terracing and canal building to make possible more efficient cultivation, which in turn produced land of such value that it could not be abandoned at the end of a few years. Families not foresighted enough to acquire such land when they had the opportunity found themselves permanently excluded from this occupation, so a landowner class developed in a previously classless society. The dispossessed continued to migrate as needed, breaking up former lineage and village units; thus, little by little, the pattern of independent village groups was transformed into a tribal one.

With capital intensive agriculture, slaves acquired a value not previously present; hence, raids to take slaves became a part of life, and slavery as an institution was introduced. Finally, about 1840, one clan established hegemony over the other settled peoples, declared itself "royal," and announced that the hereditary head of its senior lineage was now King of the Tanala. In little more than a century a new technique—irrigated wet rice cultivation—transformed the mobile, self-contained, joint family, classless village communities into the Tanala Kingdom with central authority, settled people, and rudimentary social classes based on wealth. "The transformation," says Linton, "can be traced step by step and at every step we find irrigated rice at the bottom of the change. It created a condition which necessitated either a modfication of preexisting patterns or the adoption of patterns already developed in the neighboring tribes who had had a longer time to meet these problems" (Linton 1936:353).

The same kind of transformation has occurred time after time in history, sometimes as the consequence of chance events, as in the case of the Tanala, and sometimes as the consequence of deliberately planned change. The *opposite* side of the coin, the difficulties in introducing a new technique or process because of the resistances found in the elements that touch upon it, can be illustrated by an example from the field of public health.

The keystone of community environmental sanitation is a potable water supply. During recent years sanitary engineers in many countries have designed and built thousands of water supply systems in villages and small towns previously served only by wells, streams, or lakes. Unfortunately, many of these projects have functioned at less than top efficiency: Breakdowns are frequent and repairs are slow in coming; mudholes sometimes develop around broken taps, and flies and mosquitos breed. Whole systems have fallen into disuse for one reason or another. These failures can be understood and some provision made to prevent them if it is recognized that a water supply system is not simply a problem in engineering design, but rather a function of the total way of life of a group. A successful water installation requires a whole series of preconditions; and if these preconditions do not exist, or cannot be created, failure is probable. If water in quantity is to be brought into a community, simultaneous provision must be made to carry away waste; this usually means a sewer system, or at least septic tanks or soakage pits. Trained personnel to maintain pumps and other mechanical components are essential; provision must be made for stocking spare parts in case of breakdown; transportation must be such that chemicals and other supplies will arrive regularly; an administration capable of collecting water bills or taxes is essential; the layout of water lines and outlets should conform to local architectural forms and be in harmony with the social needs and desires of the people. Users must have the will to turn water off when it is not in use rather than leaving taps running.

To the extent that these, and many other, conditions do not exist or cannot be created, a water supply system will fall short of perfection. If disposal means are lacking or if people refuse to keep taps closed when not in use, mudholes will appear and breed disease; without trained personnel, machinery will fail and the town will be without water; without spare parts and the regular arrival of chemicals and supplies, trained personnel will be unable to work efficiently; without money to maintain the system, personnel cannot be hired and supplies cannot be purchased; if outlets are planned according to the standards of other cultures, violence will be done to many social customs—women may be deprived of their social opportunities over washtubs, or youths may be deprived of courting opportunities at village wells.

3. *All Sociocultural Systems Are Constantly Changing; None Is Completely Static.* Although every society produces inventors and discoverers who are the ultimate sources of change, no group would progress rapidly if change could come about only through the ingenuity of its own members. If the opportunity for change were so limited, we should all still be in the Stone Age. As far as a particular society is concerned, its proneness to advancement is the result of its members' exposure to the tools, techniques, and ideas of other groups, their readiness to recognize advantages in ways and forms not their own, and their opportunity to accept these ways and forms, should they wish to do so. The complexity of our modern civilizations is due in only small part to the geniuses each has bred. It is due in far greater part to the willingness of our ancestors, over countless generations, to see merit in the ways of other people and to adopt these ways as their own when they saw advantage in so doing. Linton's classic passage should remind the American of his debt to world cultures and perhaps restore a bit of humility that is not always present.

> *Our solid American citizen awakens in a bed built on a pattern which originated in the Near East but which was modified in Northern Europe before it was transmitted to America. He throws back covers made from cotton domesticated in India, or linen, domesticated in the Near East, or wool from sheep, also domesticated in the Near East, or silk, the use of which was discovered in China. All of these materials have been spun and woven by processes invented in the Near East. He slips into his moccasins, invented by the Indians of the Eastern woodlands, and goes to the bathroom, whose fixtures are a mixture of European and American inventions, both of recent date. He takes off his pajamas, a garment invented in India, and washes with soap invented by the ancient Gauls. He then shaves, a masochistic rite which seems to have been derived from either Sumer or ancient Egypt.*
>
> *Returning to the bedroom, he removes his clothes from a chair of southern European type and proceeds to dress. He puts on garments whose form originally derived from the*

*skin clothing of the nomads of the Asiatic steppes, puts on
shoes made from skins tanned by a process invented in
ancient Egypt and cut to a pattern derived from the classical
civilizations of the Mediterranean, and ties around his neck
a strip of bright-colored cloth which is a vestigial survival
of the shoulder shawls worn by the seventeenth-century
Croatians. Before going out for breakfast he glances through
the window, made of glass invented in Egypt, and if it is
raining puts on overshoes made of rubber discovered by the
Central American Indians and takes an umbrella, invented
in southeastern Asia. Upon his head he puts a hat made of
felt, a material invented in the Asiatic steppes.*

*On his way to breakfast he stops to buy a paper, paying
for it with coins, an ancient Lydian invention. At the
restaurant a whole new series of borrowed elements confronts
him. His plate is made of a form of pottery invented in China.
His knife is of steel, an alloy first made in southern India,
his fork a medieval Italian invention, and his spoon a derivative
of a Roman original. He begins breakfast with an orange,
from the eastern Mediterranean, a canteloupe from Persia,
or perhaps a piece of African watermelon. With this he has
coffee, an Abyssinian plant, with cream and sugar. Both the
domestication of cows and the idea of milking them originated
in the Near East, while sugar was first made in India. After
his fruit and first coffee he goes on to waffles, cakes made
by a Scandinavian technique from wheat domesticated in
Asia Minor. Over these he pours maple syrup, invented by
Indians of the Eastern Woodlands. As a side dish he may have
the egg of a species of bird domesticated in Indo-China,
or thin strips of the flesh of an animal domesticated in Eastern
Asia which have been salted and smoked by a process
developed in northern Europe.*

*When our friend has finished eating he settles back to
smoke, an American Indian habit, consuming a plant
domesticated in Brazil in either a pipe, derived from the
Indians of Virginia, or a cigarette, derived from Mexico.
If he is hardy enough he may even attempt a cigar, transmitted
to us from the Antilles by way of Spain. While smoking he
reads the news of the day, imprinted in characters invented
by the ancient Semites upon a material invented in China
by a process invented in Germany. As he absorbs the accounts
of foreign troubles he will, if he is a good conservative*

*citizen, thank a Hebrew deity in an Indo-European language that he is 100 percent American (Linton 1936:326–327).*[3]

4. *Every Culture Has a Value System.* Although the members of a society share common values, a *value system* customarily is thought of as a part of cultural rather than of society as such. I therefore prefer to consider values in the context of cultural systems alone. All of us, to a greater or lesser extent, react emotionally to our culture. We are not neutral in our attitudes toward most of its elements. We classify the phenomena of our existence into good and bad, desirable and undesirable, right and wrong. The particular way in which we, as individuals, classify, reflects the cultural orientation of the group in which we have been socialized. Toward some things we react strongly, with approval or disapprobation. The typical American anticipates a good beefsteak with pleasure, but reacts with revulsion to the idea of eating rattlesnake meat, snails, or grasshoppers. To the Chinese a nearpetrified egg is an attractive delicacy, but the thought of drinking a glass of cow's milk fills him with horror.

A value system gives stability to a culture. It can be thought of as a balance wheel or a mechanical governor. It justifies us in our actions or thoughts and reassures us that we are behaving as our society expects. The rightness of our way of life is thereby validated. We know that behavior that significantly deviates from the norms established by our value system will be met by threats and punishment, both legal and supernatural, and that behavior that conforms to the norms will be rewarded in a variety of ways. Most individuals find security in conforming to the standards of their culture's value system. In an analytical sense a value system plays an important role in preserving a society. Values seem to change more slowly than other aspects of culture. Although this reluctance to change in the face of rapid technological advances often induces serious stresses, at the same time this essential conservatism of values serves as a brake on uncontrolled change, usually slowing the process to the point where a society can assimilate innovations without threat to its basic structure. The tenacity of value orientation is, as we shall see, one of the points to which greatest attention must be paid in planning and executing programs of technological change.

5. *Cultural forms, and the behavior of individual members of*

[3]From *The Study of Man*, by Ralph Linton. Copyright, 1936, Appleton-Century-Crofts. By permission of Appleton-Century-Crofts.

*a society, stem from, or are functions of cognitive orientations, of deep-seated premises.* Fei and Chang have written:

> Human behavior is always motivated by certain purposes, and these purposes grow out of sets of assumptions which are not usually recognized by those who hold them. The basic premises of a particular culture are unconsciously accepted by the individual through his constant and exclusive participation in that culture. It is these assumptions—the essence of all the culturally conditioned purposes, motives, and principles—which determine the behavior of a people, underlie all the institutions of a community, and give them unity (1945:81-82).

All the members of a group—a Chinese village, an American public health bureaucracy, an entire nation—share a series of common cognitive orientations, a comprehension and interpretation of the world around them, which set the terms on which they feel life is lived. I say "series of cognitive orientations" advisedly because as we progress from simpler to more complex societies, increasing numbers of levels of planes of integration are involved, and the individual belongs to more and more specialized groups, each of which shares common premises distinct from those of many other groups belonging to the same society.

Cognitive orientations therefore provide the members of a society with basic premises or postulates, which serve as guides for behavior. Some premises are fairly explicit, in that they exist at a conscious level that is recognized by thoughtful members of the group; other premises are implicit, covert, deeply buried in the unconscious. These deeply buried premises are analogous to grammar in language. Few people, while speaking, are concerned with the grammatical structures of their languages. Most of the world's people, in fact, do not even know that there is such a thing as grammar—rules and patterned regularities—underlying their speech. Yet all of us by the mere act of speaking act as if these rules were uppermost in our minds. Rules, regularities, and symbolic meanings of which we are momentarily unconscious or totally unaware determine how we speak. By the same token, most of us are totally unaware of the premises underlying our cultures and individual behavior, and certainly we do not consciously act upon them. Yet, as with speech, we behave as if these premises, these postulates, were always uppermost in our minds.

We can conceptualize the premises that characterize each of us as points along a continuum, one pole of which represents the overt, conscious level of postulate, and the other the covert, subconscious level; what is explicit for some of us may be implicit for others, and vice versa. As do Fei and Chang, I see the implicit, covert level of premise as more determinative of behavior than the explicit, overt level, in the same way that the subconscious levels of mental activity take precedence over the conscious levels in determining individual personality. The Mexican psychiatrist, Rogelio Díaz-Guerrero, who is speaking in the same terms when he describes a "sociocultural premise" as an "affirmation" that provides the basis for the specific logic of a group, seems to be in agreement:

> A sociocultural premise may be a clearly conscious
> assumption upon which a given group bases its thinking,
> feeling and behavior and it may also be unconscious, i.e.,
> a not clearly verbalized assumption which may still—or
> perhaps because of this—be even more powerful in its effects
> upon the thinking, feeling and action of the individuals of
> a given group (1967:263).

Both individual behavior and cultural forms and institutions, therefore, can be viewed as functions of, or responses to, the premises—and particularly the subconscious premises—shared by the members of a group. Premises lie at the root of our life strategies, the ways in which we construct our personal realities and organize our lives to maximize those things we want. Our premises guide us in how we relate to other people, organize our productive efforts, and use our leisure and recreation times. They underlie our feelings about religion and the supernatural, they set our logic, and they express our basic values. I believe that everything we do, everything we feel, every thought we have, is at some level tied to a validating premise.

As guides, premises may be accurate or inaccurate reflections of the conditions of life, good or bad perceptions of reality. When premises are soundly based, when they represent reasonable appraisals of reality, individual and cultural behavior forms are adaptive, realistic responses to an environment. When they are unsoundly based, behavior is less adaptive, and may perhaps be viewed as deviant, or a problem, by those members of the group who are more attuned to reality. Many of the developmental problems facing the world today stem from the fact that traditional premises valid in an earlier period now are outdated, and hence poor guides for contemporary life strategies. Yet the behavior engendered at an earlier time lingers on, discouraging innovation and encouraging traditional peoples to cling to customs that worked well in the past.

Out-of-date premises are by no means limited to traditional peoples; in greater or lesser degree they characterize all of us, for cognitive agility is never quite sufficient for the individual to be fully aware of his changing environment and to modify his behavior so as always to exploit contemporary opportunities. In technical assistance programs it is particularly important to recognize that every bureaucracy is characterized by outmoded premises that linger on, hampering efficiency and limiting action, just as the villages in which technicians work are marked by nonfunctional premises.

Chapter 2 gives an illustration of how a major premise—*limited good*—helps to explain a great deal of behavior in traditional peasant communities. Premises as they affect the work of technical specialists are further discussed in Chapter 9.

6. *Culture Makes Possible the Reasonably Efficient, largely Automatic Interaction Between Members of a Society that is a Prerequisite to Social Life.* Some of the interrelations between culture and society become clearer when we study the way in which culture, through language and other symbols, provides for the communication and understanding that is essential to the ongoing activities of daily living. Culture may be thought of as a memory bank where knowledge is stored, available immediately and usually without conscious effort to guide us in the situations in which we routinely find ourselves. Culture supplies the "tips" or "cues" that enable us to understand and anticipate the behavior of other people and to know how to respond to it.

When a person goes to live and work in a society other than his own, even though he understands the language, he will not work effectively until he learns the tips and cues that reveal the true significance of language and behavior. J. B. Adams describes how, in an Egyptian village, it is not always easy for an individual to determine at a given moment who are his friends and who are his enemies. Overt relations between people are always structured through elaborate but completely stereotyped expressions of esteem and respect that need have no bearing on the underlying feeling. When two persons exchange greetings, they look to subtle qualities of tone, pitch, and melody in speech for signs of friendliness or enmity, since the verbal expressions themselves are always the same and hence are no guide. Certain melodic patterns denote sincerity, others sarcasm, others irony, and still others hostility. Ways of accenting, abbreviating, or elongating words, and the tones in which they are uttered, convey similar meanings. "These qualities, in their different modes, are interpretable to one who is acquainted with their culturally defined meanings" (J. B. Adams 1957:226). An American unfamiliar with these cues may find himself confused when working in an Egyptian village, for the speech melody and rhythm that sound cross and belligerent to him in fact connote sincerity. Conversely, the Egyptian interprets as hostile and vengeful the melody and rhythm that to the American mean righteous indignation.

The function of culture as a shorthand device to facilitate interpersonal relations can perhaps be made more clear by returning briefly to the concepts of status and role. As previously pointed out, the term *status* refers to the positions or places of an individual in his society—not simply his social position, but all of the positions he may occupy from time to time, such as child, parent, buyer, seller, boss, worker, health educator, professor, club president, and so on. *Role* refers to the sum total of behavior patterns, including attitudes, values, and expectations associated with a particular status. For example, the status of father in American society correlates with a particular role exemplified in more-or-less stereotyped forms of behavior on the part of a man toward his children. Allowing for great individual differences, the American child learns to expect certain forms of behavior on the part of his father that are vastly different from the paternal behavior expected by a Hopi child. Conversely, the American child soon learns general forms of filial behavior that his culture tells the father he may reasonably expect from his offspring.

All of us, in the course of the day, occupy a series of statuses for which we have learned the proper role, the customary behavior patterns. When we know the behavior expected of us by our culture for each status we occupy and the behavior associated with the statuses of the people with whom we interact, we achieve a psychological security otherwise unattainable. We feel at home in the situation, we can pretty well predict what we are going to do, and we know in a general way how others will respond to our actions. It is this ability to know how to act and to predict how others will act that makes it possible for members of the same society to function together.

Let us illustrate this point with an example from the United States. As adults we purchase food almost daily, buy clothing frequently, occasionally acquire an antomobile, and perhaps once or twice in a lifetime invest in a home. In these four situations we play the role of buyer, a spender of money, but our behavior will be quite different in each situation, as will that of the seller. When purchasing food we pay the asking price without question, usually immediately and in cash, and we accept the product just as it is. When buying clothing we also pay the asking price, but perhaps insist on certain alterations as a part of the bargain. We may also reserve the right not to pay until a few days or weeks later. When shopping for an automobile we would not think of paying the asking price, and in all likelihood we will insist that the seller accept in part payment a partially worn-out specimen of the very thing we want. The remaining payments may be stretched out over a period of two or even three years. In buying a house we again do not expect to pay the asking price, and we know we can reasonably expect terms over a period of twenty or more years; unlike the example of the automobile purchase, however, we probably cannot trade in our old home in part payment on the new.

The behavior of buyer or seller that is appropriate to any one of these situations would appear ludicrous in any of the others. But since both buyer and seller have, as sharers of the same culture, learned the same rules of the game, no transaction need be preceded by lengthly palaver to establish common meeting ground.

Kluckhohn was thinking in these terms when he compared culture to a map: "If a map is accurate and you can read it, you won't get lost; if you know a culture, you will know your way around in the life of a society (1949:28–29). But, however efficient the socialization process, the fact that societies and cultures change means that no culture, however well learned, is a completely satisfactory device for preparing people to live together, just as no map can remain completely up to date. The stereotyped behavior appropriate to a particular role must change as the nature of the role itself changes. Consequently, in a rapidly changing society it is difficult to update previously learned role behavior to fit the contemporary scene exactly. To illustrate, the past generation has brought striking modifications in the roles of parents and children in the United States. The behavior patterns for children and parents learned by young adults in their childhood are only partially useful today. Today's parents must continually adjust their attitudes, make concessions, and tolerate filial behavior unacceptable in their childhood if they are to avoid complete frustration and possible breakdown.

If this is true for reasonably well adjusted members of a single society in which major changes are internally generated, think of the problems facing individuals experiencing a wide variety of strange behavior forms not indigenous to their previous way of life. The immigrant father in the United States, for example, is confronted not only with the evolutionary changes in parent-child role behavior of his own culture, but also with the very much greater differences he finds in his adopted country. Frequently he has been brought up to feel his role requires authoritarian attitudes that are at variance with modern American practice. He expects degrees of obedience and respect normally absent in today's children; and when the behavior of his offspring, conditioned by that of their schoolmates, does not conform to his expectations, his security is threatened, he feels his children are running wild, and he functions less effectively as a father than he or his children wish.

Disparate conceptions of appropriate role behavior characterize many of the situations in which planners and practitioners of directed culture-change work. The patient in a peasant community has a vastly different idea of the correct comportment of the curer than does the individual who since childhood has been familiar with scientific medical practice. And when the physician finds himself in a setting where his patients' concepts of respective roles and associated behavior are very different from his own, he will experience some of the same sense of bewilderment and frustration as does the immigrant father in the United States. The practice of medicine in such situations is carried on under considerable handicap.

Today's peasant farmers often find themselves in similar, and in some ways more extreme, situations. Not only do they have no idea of the appropriate behavior associated with the role of the agricultural extension agent; they do not know, until the agent arrives on the scene, that there is such a role. What they have learned is that the role of traditional government officials, vis-à-vis their welfare, has been prejudicial to their interests. It takes some time before peasant farmers can develop faith in a different type of relationship with this new government official, who says he has come to help and not to take away.

In subsequent chapters the strategic importance of a clear understanding and perception of role and role behavior by both initiators and recipients of programs will become apparent.

# 2

## the rural community: the traditional world

### The pastoral ideal

Since the end of World War II a large part of technological assistance has been directed toward rural communities. This is not surprising, for despite the rapid urbanization of the twentieth century more than half the world's population still dwells in villages. Almost always the inhabitants of these villages have been marginal, if not in a geographical sense, certainly in a cultural sense. Lacking education, communication, and knowledge of the wider world, they have lagged behind their city cousins in almost all aspects of economic and social development. It is only just, then, that specialists in public health, education, agriculture, and community development should have devoted special attention to the needs of rural peoples.

It is axiomatic that, in order to work successfully among a group of people, it is important to know a good deal about their cultural and social forms. So we ask the question, "How well have American (and international) technical specialists understood the rural societies in which they have worked?" Unfortunately the answer often is, "Not nearly well enough." The reasons for poor understanding are twofold. First, as the anthropologist testifies, it takes specialized training and much time and hard work to delve into the inner workings of a strange society, and only recently have we begun to accumulate really good information about the nature, range, and variety of the world's rural peoples. And, second, Western man has inherited a preconceived idea of the nature and quality of rural life that he resolutely refuses to abandon in the face of mounting evidence to the contrary. I refer to the rural-life-is-ideal theme, which, under labels such as the "pastoral ideal," the "agrarian myth," the "arcadian myth," and the "cult of the noble savage," can be traced back as far as classical antiquity. For as long as there is any record, Western man has believed, and wanted to believe, that rural life is marked by an especially moral quality, that it encapsulates the fundamental virtues of his society, standing in opposition to the city, with its impersonality, disorganization, and *anomie,* its vices and competitive pressures. This is one of our oldest and most cherished beliefs, and to suggest that it may not be true is seen by some as an attack on the very foundations of contemporary society. Man appears to have felt guilty for the invention of the city and, to clear his conscience, he apologizes for his deed by insisting that the basic goodness of the preurban world still survives undiminished in rural communities.

Both Caro Baroja and Susan Lowenstein, in independent essays, have pointed out how, since the time of ancient Greece, urban moralists and intellectuals have deplored the real and imagined vices of city life, and have searched for, and sometimes found, a lost Utopia in the "natural" and unspoiled ways of the traditional rural community. When Aristophanes (448?–?380 B.C.) in *The Clouds* portrayed the difference between the good life of old and the contemporary life of the city, he took as prototype of the former Strepsiades, a pious, industrious man who loved country life, and as prototypes of the latter his wife and son, city people "addicted to all sorts of diversions, extravagance and superfluities" (Caro Baroja 1963:27–28). The Roman scholar and agronomist Varro (116–?27 B.C.), in his famous treatise on agriculture, recognized two forms of life: urban and rural. The latter, he said, is not only anterior to the former, "but it is also more noble and better, since it is given to us by divine nature, whereas urban life is the work of man" (Caro Baroja 1963:27).

Aristophanes, says Caro Baroja, was repeating the common view of Athenians of his time. "Yet this commonplace has lasted to our day; in the city are found vice, corruption and artifice; in the country the ancient virtues, and still more than in the countryside of one's own land, in the countryside of distant regions which have a smaller number of cities" (Caro Baroja 1963:28).

The Romans inherited this dichotomized view of rural and urban life. Cato, writing in the century before Christ, expressed the belief that "it is from the tillers of the soil that spring the best citizens, the stanchest soldiers; and theirs are the enduring rewards which are most grateful and least envied. Such as devote themselves to that pursuit are least of all men given to evil counsels" (Lowenstein 1965:118). Columella, in the first century A.D., held similar views: Those people "who spent their time idly within the walls, in the shelter of the city, were looked upon as more sluggish than those who tilled the fields or supervised the labours of the tillers" (Lowenstein 1965:114). His contemporary, the satirical Roman poet Juvenal (ca. A.D. 60–ca. 140), saw Rome as "a great sewer," reserving his praise for the unspoiled qualities of country people. Cicero concurred: "Personally I incline to the opinion that no life could be happier than the farmer's," and, the "pleasures of farming . . . come closest of all things to a life of true wisdom" (Lowenstein 1965:118).

Through succeeding centuries such writers as Rousseau, Thoreau, Wordsworth, and Emerson repeat this idealized view of rural life. It was implicit in the thoughts of our founding fathers, perhaps most clearly stated in Jefferson's earlier writings, where he views farms not primarily as productive economic units but rather as a means of preserving rural manners and rural virtues. So imbued was he with the idea of the evil city that he advocated shipping American raw products to Europe for manufacture and then reimporting them. The loss of income, he argued, would be more than offset by happiness and permanence of government (Marx 1964:126–127). "State a moral case to a ploughman and a professor," said Jefferson. "The former will decide it as well, and often better than the latter, because he has not been led astray by artificial rules" (Marx 1964:130).[1]

The *pastoral ideal* has since the last century underlain American agricultural policy, which rests on the assumption that the single-family farm must at all costs be maintained if the United States is to continue as a healthy democracy.[2] And the pastoral ideal has been

[1]In passing, it may be noted that praise for rural virtues and rural life generally emanate from urban intellectuals of substantial means not faced with the reality of grubbing a marginal existence from the scanty resources of a peasant village.
[2]The United States is not alone in linking the preservation of the pastoral ideal with rural policy. In describing the Castillian village where he lived, Aceves writes: "One of the most pervasive ideologies is the so-called Arcadian Myth. It is illustrated by governmental attempts to industralize Spain without having

implicit in the thinking of American and other technicians—many of them from rural backgrounds—who have worked in technical-assistance programs. Their strategies for rural work tend to be based on the assumption that villages are inherently well integrated, and that villagers are by nature, and in contrast to urban dwellers, basically cooperative. To the extent that they have found otherwise, they have assumed that recent urban contamination has infected a formerly sound organism, so that their task is to restore the village to its former "natural" state and then to build on its virtues and strengths.

### Peasant character

If the pastoral-ideal model seriously misrepresents the quality of rural life, what then is reality, and how can we discover it? The first thing we note is that even in classical antiquity not all writers agreed about the perfection of rural life. Hesiod (eighth century B.C.), who is much quoted by those who subscribe to the pastoral ideal, cautions his readers: "When you deal with your brother, be pleasant, but get a witness: for too much trustfulness, and too much suspicion, have proved men's undoings" (Hesiod 1959:63). The Old Testament, too, takes note of man's shortcomings: "It is better to take refuge in the Lord than to put confidence in man" (Psalms 118:8).

It is largely to the work of anthropologists, however, that we must turn to learn about rural life. Even they have been late in acknowledging that the pastoral-ideal model does not fit, a delay perhaps explicable as an historical accident. When the so-called Chicago school of urban sociologists—Robert Park, Louis Wirth, and Ernest Burgess—began their epoch-making urban research in Chicago during the first third of the present century, they appear to have approached the city with a rural bias, finding in it great social pathology. The anthropologist Robert Redfield was a student at the University of Chicago during this period of intellectual ferment; he was also the son-in-law of Park. Not surprisingly, when he went to Mexico to carry out his doctoral research, he took with him the pastoral ideal as a model of what he would find. *Tepoztlán*, the first major peasant study by an American anthropologist, and one of the

---

peasants leave the farms. The rationalizations given to support the Arcadian Myth are many and varied, but their underlying rationale—if the term may be used here—is that the small towns, the *mini-pueblos*, are sort of a national 'moral reserve,' and their inhabitants are sort of Noble Savages living in an idyllic environment away from the vice and corruption of the city" (Aceves 1971:107–108).

first to be made by any anthropologist, portrayed the village of this name in terms sufficiently idyllic to delight the most jaded city dweller (Redfield 1930). Here, for the first time, man's preconceptions of rural life were verified by scientific research. Redfield subsequently elaborated his views in a series of books and papers, which culminated in "The Folk Society" (Redfield 1947), in which he described peasant and primitive societies as a type contrasting to urban city societies. In this paper it is clear that Redfield, like many urban dwellers before him, is drawn emotionally to the qualities of life he finds in the simpler societies.

Redfield's writings attracted attention far beyond anthropology, for he was justly recognized as the humanist-philosopher-anthropologist of his time. Social scientists were aghast, then, when Oscar Lewis, after an intensive follow-up study in Tepoztlán in the 1940s, came to very different conclusions about the quality of life in the village. In contrast to the homogeneous, well-integrated, smoothly functioning community of happy and well-adjusted people that Redfield described, Lewis was struck by "the lack of cooperation, the tensions between villages within the municipio, the schisms within the village, and the pervading quality of fear, envy, and distrust in inter-personal relations" (Lewis 1951:429).

> Gossip is unrelenting and harsh in Tepoztlán. . . . Facts
> about people are unconsciously or maliciously distorted. . . .
> Relatives and neighbors are quick to believe the worst,
> and motives are always under question. . . . Successful persons
> are popular targets of criticism, envy, and malicious gossip
> (Lewis 1951:294).

Had Lewis directly attacked motherhood, he could hardly have raised a greater storm, for the destruction of a sacred myth is the gravest of all threats to any society. Yet time has proved Lewis to be correct in his basic interpretation, not just of Tepoztlán, but of peasant societies in general.

Most anthropologists, and the rarer sociologist, psychologist, and political scientist, who have lived in peasant communities are in general agreement about the nature of peasant society and character. Usually they have been genuinely fond of the people who have accepted them into their villages. After overcoming initial suspicion they have found warm friendship, a zest for life, humor, a spirit of rugged independence, and other attractive personal qualities. They have also, most of them, found much criticism of fellow villagers, suspicion, distrust, envy, and lack of cooperation in settings other than those in which very careful reciprocity rules have been worked out.

In other words, peasants are like most other people: They have some very nice qualities, but they also display character traits that can hardly be said to conform to the pastoral-ideal model. While anthropologists believe that traditional peasant behavior is a realistic response to the harsh conditions that historically have characterized their lives, they also see that much of this behavior is ill-adapted to new opportunities found in the contemporary world. Our problem, therefore, is to account for this traditional behavior, so that we can look for ways in which the peasant can be helped to modify his behavior in the direction of new attitudes and practices better suited to the present.

### Peasant society

What is distinctive about peasant society? How can we define peasant communities? When we raise such questions, the first thing that strikes us is that peasant societies differ from other small rural groups such as historic North American Indian tribes, the Australian bushmen, and Polynesian islanders in that they are not isolated, essentially self-sufficient units. Whereas these "primitive" societies, and hundreds of other similar ones, have all been self-sufficient with respect to their religion, their philosophy, and their government, most peasants depend on the city and state for this kind of nourishment. The peasants' religion is usually a simplified form of what Redfield called the "great tradition" of their larger society; it comes to them from without, and is not an autochthonous creation. Many of the peasants' values likewise represent diffusion from urban centers. Politically, peasants have little independence; For longer than they can remember they have been governed by the state, and as a consequence their own communities have but weakly developed leadership patterns.

It is clear that peasants exist in an intimate relationship with towns, cities, and the state. As Kroeber long ago put it, "They form a class segment of a larger population which usually contains also urban centers. . . . They constitute part-societies with part-cultures" (Kroeber 1948:284). Peasants are primarily farmers, but they may be artisans and fishermen as well. They produce much of their food, and they are able to make many of the material items they need, such as clothing and tools. But they depend on town markets to sell surplus produce and to buy items they cannot make or grow themselves. Thus, although peasants are primarily agricultural, the criteria of definition are structural and relational rather than occupational. It is not *what* peasants produce that makes them peasants; it is *how* they produce and to *whom* they dispose of what they produce, as well as to whom they are politically subservient, that is determinative. Peasant communities represent the rural expression of large, class-structured, economically complex, preindustrial civilizations, in which trade and commerce and craft specialization are well developed, in which money is commonly used, and in which market disposition is the goal for a part of the producer's efforts. The city is the principal source of innovation for such communities; it, and the state, hold the political, religious, and economic reins.

The rural peoples who wholly or in large part conform to these criteria include the mestizo and settled Indian-language-speaking villagers of Latin America; those who surround the Mediterranean in Europe, Asia, and Africa; and peoples of the Middle East and South and East Asia. African village dwellers south of the Sahara are viewed by most anthropologists as semipeasants, for although most of them historically are settled horticulturalists, the nexus with city life and the state usually is less marked, and often altogether lacking (Fallers 1961).

The most important implication of this view of peasant society is that the peasant is virtually powerless with respect to large areas of his life, because the basic decisions affecting him are made by members of other classes. Political activity is truncated, because major control is exercised from national or provincial centers. Economically, the peasant is dependent on forces that operate well beyond his local boundaries, and only under special circumstances are prices for his production set by village factors. There is not even local autonomy in religion, at least not in Latin America and the Mediterranean, for obligations and observances are set and guided by doctrine and tradition that far transcend national boundaries. In other words, the peasant expects to obey, not to command. For generations he has been able to show initiative only in the most limited areas. Small wonder that he often has trouble in making up his mind about something new. Moreover, not only does the peasant have little or no control over the basic decisions made from the outside, but *usually he doesn't even know how or why they are made.* The orders, the levies, the restrictions, the taxes that are imposed from the outside have for him the same quality of chance and capriciousness as do the visitations of the supernatural world. And the peasant feels much the same toward both the authorities of the city and the supernatural: He can plead, implore, propitiate, and hope for a miracle, but in neither case can he expect by his own action to have any effective control. A fatalistic attitude toward life? It is hard to imagine more favorable circumstances in which it could develop, and hard to understand how it can be lessened until these circumstances are changed.

The peasant has been victimized by persons more knowledgeable than he since the beginning of time. He knows he is a rustic, a country bumpkin who, in his necessary trips to town, will be taken advantage of by men without conscience. He needs the city, but he hates and fears it. He fumes at humiliation and imagines slights even when they are not intended. The Wisers, in the last chapter of their wonderful little book, *Behind Mud Walls*, sensitively put into words the Hindu villagers' feelings about their helplessness.

> *In the cities they devise ways of exploiting us. . . .*
> *When we get our money and want to take home some cloth,*
> *the shopkeepers get out the pieces which they have*
> *been unable to dispose of, and persuade us to buy them at*
> *exorbitant prices. We know that they are laughing at us.*
> *But we want cloth, and the next shopkeeper will cheat*
> *us as badly as the last. Wherever we go in the town, sharp eyes*
> *are watching to tempt our precious rupees from us. And*
> *there is no one to advise us honestly or to help us escape from*
> *fraudulent men.*
>
> *You cannot know, unless you are a villager, how everyone*
> *threatens and takes from us. When you [the Wisers] go*
> *anywhere, or when a sophisticated town man goes*
> *anywhere, he demands service and gets it. We stand dumb and*
> *show our fear and they trample on us (Wiser and Wiser*
> *1951:163, 167).*

In the mestizo village of Tzintzuntzan, Mexico, in which I have done research for many years, I know an intelligent and imaginative young man, a pottery merchant. He buys the lovely ware made by his fellow villagers and delivers it on commission to stores in Mexico City. At least once a month he climbs into the bus for the overnight run to the capital, his crates of pottery strapped to the roof. Upon his arrival he makes his deliveries, admires the city, and climbs on the bus for the return trip. The bus is like home; it is known, and safe. He has never spent a night in the big city, and he tells me he wouldn't know how to go about it.

Another villager in Tzintzuntzan, recognizing a need for better bus service to the nearby market town of Pátzcuaro, bought a secondhand bus from an official of a local line. He was delighted, and so were his neighbors. Now it would not be necessary to wait on the highway while bus after loaded bus passed without stopping; Tzintzuntzan would have its own service. Then our entrepreneur began learning about the outside world. A franchise? Bus drivers unions? The official who sold the bus fought him on both scores, and won. Sadly he sold the bus back to the same man, at a fraction of its cost to him.

The peasant has learned that the outside world is fraught with dangers, that it is unpredictable and cannot be understood. Is it surprising that he has come to value his traditional ways and the predictable quality of life within this microcosmic world, his village?

With the outside world a constant threat, it might seem logical to expect a high degree of cooperation among peasants as a defense mechanism if for no other reason. Should not common adversity draw people together in the pursuit of common ends? Cannot a villager find strength in unity? However natural this might seem, and however frequently it is assumed by the technical specialist working with peasant villagers, the evidence, as we have seen, indicates that this is not so. Lewis's findings in Tepoztlán have been corroborated in a growing list of studies from all parts of the world.

Friedmann, in his superb analysis of the "world of *La Miseria*" in Calabria and Lucania, dwells upon the peasants' "mentality of mutual distrust," their inability to work cooperatively and to collaborate for the common good; *La Miseria* is a "world in which to love one's neighbors, to let down one's guard in the face of the relentless struggle for existence, would simply mean to commit suicide" (Friedmann 1958:21, 24). Banfield found, in the south Italian village of Montegrano, that friends and neighbors are considered potentially dangerous. No family can stand to see another prosper without feeling envy and wishing the other harm; beyond the nuclear family, he finds, concerted action for the common good is impossible (Banfield 1958:10, 121).

The Spanish picture is much the same. Aceves writes that "the Castillian peasant frequently views his world as a place where trouble lurks at every turn; a world of suspicions and mistrust where order is a tenuous thing difficult to achieve and more difficult to maintain"

(Aceves 1971:127). Fears of disorder, disharmony, and trouble are always present, as evidenced by barred windows, secrecy in business details, "and the ever-present fear of the ubiquitous 'they' who seem to be so powerful an agent of social control" (Aceves 1971:127).

The physician-psychiatrist Carstairs found a similar picture in a Rajasthan village in India. Villagers often made enthusiastic plans to work together for the mutual good, but these plans were rarely carried out. "Within an hour or two, one of the group would warn me that someone else was only in the scheme for his own advantage. . . . From the beginning to the end of my stay, my notebooks record instances of suspicion and mutual distrust" (Carstairs 1958:40).

The list of illustrations could be extended, but the point should be clear: To a greater or lesser extent peasant life is characterized by, in addition to the "desirable" qualities we have noted, a bitter quality of mutual suspicion and distrust, which makes it extremely difficult for people to cooperate for the common good. With this mentality, a new technical aid program that presupposes a high degree of village cooperation is obviously headed for trouble.

Why is the quality of village interpersonal relations often so poor? If we dig into the reasons, we find that quarrels about property, and particularly land ownership, are most frequently mentioned. For example, Simmons writes that in the Peruvian south coastal village of Lunahuaná the characteristic reaction to even those one knows well is suspicion and distrust; everyone is presumed to be out for himself and expected to use the most unscrupulous methods in pursuing his self-interest. "Sibling relations," he says, "after the abdication or death of parents, are characterized by disputes and feuds arising from conflicts over the division of land and the other goods of inheritance" (Simmons 1959:104). The Italian novelist Ignazio Silone describes village life in Fontamara in this fashion: "In bad weather months they arranged family affairs. That is, they quarreled about them. . . . Always the same squabbles, endless squabbles, passed down from generation to generation in endless lawsuits, in endless paying of fees, all to decide who owns some thornbush or other" (Silone 1934:ix). Adamic describes how the "seemingly perfect village life" of the Slovenes every once in a while is "shaken by fierce quarrels among peasants over the possession of a few feet of ground or a tool or a beast" (Adamic 1934:97). And Smith tells how in China the division of land following the death of the owner almost always results in bitter feuds, even between brothers (Smith 1889:328).

If we look beneath these overt expressions of competition and bitterness, we see a second basic characteristic of peasant life (the first is impotence in the face of the outside world) which makes behavior more intelligible. This is, *peasant economy is essentially nonproductive*; peasants ordinarily are very poor people. Their resources, particularly land, usually are absolutely limited, and there is not enough to go around. Productive techniques, based on human and animal power and the simple tools first used before the time of Christ, are essentially static. Consequently, production is constant (except as affected by weather), and perhaps it declines over the centuries, as the result of erosion, deforestation, and other consequences of man's exploitive activities. That is, the total "productive pie" of the village does not greatly change, and moreover *there is no way to increase it however hard the individual works* unless new land and improved techniques become available. The consequences of this situation are apparent: *If some one is seen to get ahead, logically it can only be at the expense of others in the village.*

## The image of limited good

The recognition of this truth led me to attempt to formulate a more comprehensive model to explain peasant behavior, a model in which not only economic good, but almost all other kinds of good—friendship, love, manliness, honor, respect, health, power, and security—are seen as severely limited (Foster 1965). The model of *limited good*, which, it must be remembered, is an ideal type and not a description of any specific society, embraces five points:

1. Peasants widely share an implicit premise, a cognitive orientation in which they perceive their socioeconomic and natural environments to constitute a closed system.
2. The system's resources—natural, economic, and human—are insufficient to meet each member's needs, to provide the good that he wants. Not only is good limited, it is finite, static, and unexpandable within the system.
3. Although peasants believe the good within their system is finite, they also know that there is more good beyond the boundaries of their system, and hence not normally available to them.
4. In a closed, static, unexpanding, and unexpandable system, as the zero-sum game model predicts, one person's gain with respect to any good must be another's loss.
5. To guard against being a loser, peasants in traditional com-

munities have developed an egalitarian, shared-poverty, equilibrium, status-quo style of life, in which by means of overt behavior and symbolic action people are discouraged from attempting major changes in their economic and other statuses.

A great deal of peasant behavior appears to be a function of this particular view of the world. A limited-good premise helps us to understand why the successful person invites the suspicion, the enmity, the gossip, the character assassination, and perhaps the witchcraft and physical attacks of his fellows. Any evidence of a change for the better in his situation is proof of guilt—it is all that is needed to show that, in some fashion ,he has taken advantage of his neighbors. The villagers not unnaturally react in the most effective way known to them to discourage a neighbor from tampering with the traditional division of the pie. The force of public opinion in peasant society, through its very bitterness and mercilessness, is thus seen as a functional device whereby families protect themselves from economic and other loss through the real or imagined chicanery and dishonesty of their friends.

This focus also explains why peasant families usually attempt to conceal their economic improvements. Visible evidence of fortune will be interpreted as an open confession of guilt, and the lucky or hardworking family will be subject to slander and gossip and perhaps economic blackmail as a consequence. Again it is the Wisers who make so clear the villagers' attitude toward display of wealth.

> *Our walls which conceal all that we treasure, are a necessary part of our defence. . . . they are needed against those ruthless ones who come to extort. . . . our fathers built them strong enough to shut out the enemy, and made them of earth so that they might be inconspicuous. . . . But they are a better protection if instead of being kept strong they are allowed to become dilapidated. Dilapidation makes it harder for the coveteous visitor to tell who is actually poor and who simulates poverty. When men become so strong that the agents of authority work with them for their mutual benefit, they dare to expose their prosperity in walls of better materials and workmanship. But if the ordinary man suddenly makes his wall conspicuous, the extortioner is on his trail. You remember what a short time it was after Puri put up his imposing new verandah with a good grass roof, that*

*the police watchman threatened to bring a false charge*
*against him. He paid well for his show of progress. Old walls*
*tell no tales (Wiser and Wiser 1951:157).*

In many peasant communities wealth can be displayed only in a ritual context—a church fiesta, for example—in which the pious individual may in fact go deeply into debt. So we have the striking contrast of poverty-stricken people spending enormous (for them) sums of money to maintain their standing in the community. And the individual who does not, at least once in his life, carry such a religious obligation, will be scorned by the tradition minded. It is almost as if the society requires him to say, "Look, I have in no way consciously improved my position at your expense, and if I have had advantages about which I have not known, I want no part of them. To prove my sincerity, I am giving away most of what I have so that you can feast and enjoy yourselves, and we can worship our God. You will see that I do not even want to keep what is rightfully mine." Anthropologists call these expensive fiestas "siphon" or "redistributive" mechanisms, which serve, symbolically or in fact, to maintain an image of relative equality within the community.

Since villagers are suspicious of the intentions of others, they try to minimize the situations in which they may be exploited by others. This they do by avoiding close ties, unless validated by ritual safeguards such as the *compadrazgo* godparent system or other forms of ritual friendship, or by permitting friendship ties to become strong only after they are very certain of the motives of others. This explains why cooperation is often so difficult to achieve in peasant communities. In an inherently distrustful society the individual is confronted with a prisoner's-dilemma payoff matrix in which cooperative urges are discouraged by the recognition that such action is apt to be exploited by the other player, to ego's detriment. Better not to risk this outcome at all, than naïvely to be drawn into a cooperative venture where one cannot control the outcome.

A limited-good model also helps us to understand the frequent peasant difficulty in developing strong local leadership. Not only are restraints placed from outside, but restraints are also placed from within. With the total village quantity of power and influence seen as limited, to delegate a part of the very little bit that the individual has to a local leader is seen as unnecessarily strengthening a potential opponent who may use that power for his own purposes. In other words, an ideal man, by merely accepting office and the authority that goes with it, is seen as depriving others of something that they value, but that they have in very limited quantity; the leader ceases to be an ideal man as soon as he assumes leadership.

A significant part of the life strategy of peasants in what are (to them) closed systems consists of trying to tap good lying outside the system. The frequent stories that explain a significant increase in well-being as the result of finding buried treasure, having made a pact with the devil, or winning on the lottery are manifestations of this view of strategy. Human and supernatural patrons from outside the system are viewed as immensely valuable. And finally, in the contemporary world, working for wages outside the system, in cities or in foreign countries, is seen as tapping outside wealth and other desirable forms of good. Since all these "goods" are known to come from beyond the system's boundaries, and since they are acknowledged as not being at the expense of others, they are "safe" and can therefore be displayed.

It is important to stress that limited-good behavior is by no means exclusive to peasant society; in some degree it is probably found in all societies. It is perhaps more accurate to say that we are faced with a continuum of good, from more limited to less limited. Members of some societies, and especially peasants, exhibit a great deal of behavior clustered toward the limited-good end of the continuum; others, including contemporary United States, exhibit more behavior that falls on the continuum toward the unlimited-good pole. By the same token, there is clearly much variation within peasant and other rural societies: Some see good as more limited than others. Brokensha and Hodge, in discussing this variation, correctly stress the distinction between *subsistence economies* and *peasant economies*. By the former they mean communities (such as those in black Africa) that produce all or most of their food and other needs, and sell little or nothing in formal markets; these stand in contrast to peasant economies in which, as we have seen, a significant amount of the villagers' production is disposed of through barter or sale. "In Africa," they write, "where there are few communities which have all of the peasant characteristics, these difficulties [i.e., behavior stemming from a limited-good outlook] do not affect the community-development process as strongly as elsewhere" (Brokensha and Hodge 1969:187). In suggesting the limited-good model to explain a great deal of rural behavior I am obviously painting with a broad brush that glosses over very real differences. I must emphasize also that the model has been constructed to explain behavior in *traditional* peasant societies, and not in contemporary, rapidly changing communities where real behavior increasingly deviates from that predicted by the model.

A full-blown limited-good premise certainly discourages progress and inhibits the efforts of technicians interested in promoting change. But it is by no means an absolute barrier; when rural peoples perceive concrete evidence of new opportunities, they are able to shed at least some of their traditional premises in remarkably short order. Once the threshold to modernity is crossed by a few innovative souls, an increasing flow of followers crosses over, and in time the limited-good premise will lose most or all of its inhibiting power.

Peasant culture places heavy burdens on those who wish to live by its rules. Yet it also provides the devices whereby, within the traditional framework, the individual is able to reach out and achieve some degree of security, however modest the level may be. Opposing the centrifugal forces that are constantly tearing at peasant societies are centripetal forces that hold them together. In some places —the Indian Village is a good example—a strong feeling of unity marks the extended family and the caste, and mutual and reciprocal obligations mark the behavior of people bound together in such units. The Wisers, again paraphrasing the villagers, write:

> No villager thinks of himself apart from his family. He
> rises or falls with it. . . . we need the strength of the
> family to support us. . . . That man is to be pitied who must
> stand alone against the dangers, seen and unseen, which beset
> him. Our families are our insurance. When a man falls ill,
> he knows that his family will care for him and his children until
> he is able to earn again. And they will be cared for
> without a word of reproach. If a man dies, his widow and
> children are sure of the protection of a home (Wiser and Wiser
> 1951:160).

Sometimes the *ideal* of familial unity is extended to the village, even reaching across castes and in the face of real schisms and divisions that make village cooperation difficult. McCormack describes how in a Mysore village "the ideal of village unity emerges as an important element in the villagers' own interpretation of proposals for village improvement," and it is strongly felt that a unanimous vote is a precondition to going ahead (McCormack 1957:257). In spite of the real picture, conflicts and notorious examples of lack of cooperation in families and within village groups are considered as "shameful," and an important task of hereditary leaders is the arbitration of disputes to restore what is imagined to be a normal village unity.

In other parts of the world—the Italian and Latin American data come to mind—the extended family is less strong, but its place is taken by a godparenthood system, tying individuals and families together with sacred sanctions that give them something of the group strength and social security afforded by an effective family unit. Formalized friendship, too, often functions to tie together the loose strands occasioned by the divisive forces extant in peasant society. In both formalized friendship patterns and the godparenthood system, we see the element of individual reciprocity based on a sense of *contract*, in contrast to the *ascriptional* basis of the kin-based group. The prevalence of these and similar types of fictive kinship (the anthropological expression for godparent, blood brother, best friend, and similar kin-like relationships) in peasant society helps to explain one of the most important things an outsider must understand in working in such a community. A contract is a bargain between two or more people. It is *personal*, entered into freely by the participants. It implies both obligations and expectations. One must help a friend or a co-godparent (a *compadre*, to use the Spanish term) when such help is requested, and to the best of one's ability. Conversely, aid and succor may be expected when needed. Relationships are *personalistic* in a society in which the individual contract is the only effective nonfamilial basis for social intercourse. One achieves through having the right friends with effective obligations toward one; conversely, one often does things for these friends, even if the reasons are not fully understood. Not infrequently community development workers have found that villagers have cooperated with them, not because they understood or desired the innovation, but because they felt they had established a friendship relation with the outsider which required that they do what the new friend asked, in return for previous or subsequent favors.

It is for reasons like this that an effective worker must thoroughly understand the social structure of a community, the forces that divide people, and the forces that draw them together. The behavior of a peasant villager, however stubborn and unreasoned it may seem to an outsider, is the product of centuries of experience. It is an effective protective device in a relatively unchanging world. It is less effective in a rapidly industrializing world, and ultimately it becomes a serious hindrance. But the peasant is pragmatic; he is not going to discard the clothing that has served him well until he is convinced that he will profit by so doing. He sees that the future holds new things, but he remembers the past:

> Our lives are oppressed by many fears. We fear the rent
> collector, we fear the police watchman, we fear everyone who

*looks as though he might claim some authority over us,*
*we fear our creditors, we fear our patrons, we fear too much*
*rain, we fear locusts, we fear thieves, we fear the evil spirits*
*which threaten our children and our animals, and we fear*
*the strength of our neighbour (Wiser and Wiser 1951:160).*

It is against the background described in this chapter that the processes of social and cultural change will be described.

# 3

## the rural community: the contemporary world

In the preceding chapter the rural community was described and interpreted as an ideal type, as it has existed for hundreds, and even thousands, of years in a preindustrial or semindustrialized world. In Europe the model began to crumble early in the nineteenth century with the development of large cities and major industry, which profoundly affected preexisting rural forms, while in the United States the type of community described can hardly be said ever to have existed. But outside of northern Europe, North America, and Australasia the traditional rural community, in its classical peasant or African horticultural variants, has existed well into the present century. Only since the end of World War II—with independence for former European colonies, with the burgeoning of small towns into huge cities, with increases in commerce and the beginnings of major industry— has the traditional community begun to experience major changes. The sources of these changes are varied, but they all add up to one profound reality: For the first time in history average villagers have options and these options are being increasingly exercised. With relative ease, they can choose *not* to be peasants, *not* to be villagers, and increasing millions are making this choice. Villagers can migrate to the large towns and cities of their countries and enter industry and the service trades. They can also choose to be semi-villagers, maintaining rural ties but earning all or a large part of their income away from home. Millions of villagers have left their countries to seek work in agriculture (to illustrate, Mexican *braceros* in the United States), in industry (workers from southern Europe in heavily industrialized north Europe), and in mines (Africans in South Africa, Zambia, and the former Congo). Villagers can also remain at home and still exercise options: Successful revolutions, major and minor, while not the rule, have been sufficiently frequent to indicate that under the right circumstances villagers can assume a degree of control over their lives that has historically been denied them.

Partially as a consequence of the exercise of options, and partially as a consequence of national independence (or, for countries already independent politically, as a consequence of increasing economic independence), traditional villagers are experiencing a second major metamorphosis: More and more they see themselves, not in local or regional terms, but as citizens of nations, as sharing history, customs, values, and expectations with millions of others about whom, until very recently, they knew little or nothing. Nationalism, and the awakened pride in nationality that goes with it, increasingly is manifest at the village level in developing countries. In Mexico, for example, hundreds of thousands of villagers, many in organized school tours, visit Mexico City annually. There, at the National Museum of Anthropology, they see visible evidence of the antiquity and richness of their indigenous civilization, and from the streets, the subway, and the airport they see exciting proof of what they hear on radio and see on television: They are citizens of a great and powerful country. And they can return home filled with pride and pleasure, aware that their interests now transcend their *patria chica,* their local homeland.

As recently as a generation ago it was possible to see rural technical assistance largely in local, agrarian terms, and most aid was planned within this framework. Local customs, local resources, local leadership, local communities, constituted the conceptual framework for a great deal of community developmental work. Although the conception was deficient even then, at least a partial case could be made for this approach. The technical specialist could work fairly well without understanding national aspirations, national development, or the extravillage ties of many of the members of the families with whom he worked. Today this is no longer possible. Successful village-level work presupposes knowledge not only of the traditional community and its world view, but also of how that com-

munity and world view are being transformed. In this chapter urbanization and nationalism, two of the major forces transforming village
life, are examined in the context of what these processes mean to
both villagers and village-level technical assistance workers.

## URBANIZATION

Anthropologists customarily use the term *urbanization* in a different
sense from that of economists and demographers. Looking at the
phenomenon from the rural end of the village–city continuum, they
see it as encompassing the whole sequence of behavior change beginning with a villager's decision to leave his community: his route
to the city, the ways in which he finds a toehold in the city, acquires
a home and a job, learns a "new" urban culture, and how he finally
becomes a city dweller fully enculturated into the ways of urban life.
In contrast, economists and demographers, who carry out their studies at the urban end of the continuum, usually define urbanization in
a statistical sense, as the percentage of the total population resident
in towns and cities. These contrasting definitions may cause confu-
sion and disagreement. Whereas to the anthropologist increasing
rural–urban migration means greater urbanization, to the economist
and demographer the index of urbanization increases only if cities
grow more rapidly than rural areas, that is, only if the proportion of
urban dwellers to rural dwellers grows. With respect to the term *urbanism* there is less disagreement: Most scholars agree with Wirth,
who defines it as "that complex of traits which makes up the characteristic mode of life in cities" (Wirth 1938:7)—that is, as urban
culture.

Many—perhaps most—urban studies, other than those made by
anthropologists, incorporate a value judgment. Unconscious prisoners of the pastoral ideal, students of social problems in urban settings
tend to contrast the obvious problems of the city with the imagined
virtues of the countryside, to arrive at an evaluation of the quality of
city life that is far from flattering. Probably no scholar has had more
impact in propagating this view than Louis Wirth, who wrote:

> The superficiality, the anonymity, and the transitory
> character of urban-social relations make intelligible. . . . the
> sophistication and the  rationality generally ascribed to
> city-dwellers. Our acquaintances tend to stand in a relationship
> of utility to us in the sense that the role which each one
> plays in our life is overwhelmingly regarded as a means for

*the achievement of our ends. Whereas, therefore, the
individual gains, on the one hand, a certain degree of
emancipation or freedom from the personal and emotional
controls of intimate groups, he loses, on the other hand, the
spontaneous self-expression, the morale, and the sense of
participation that comes from living in an integrated society.
This constitutes essentially the state of anomie or the
social void to which Durkheim alludes in attempting to account
for the various forms of social disorganization in technological
society. . . . The close living together and working
together of individuals who have no sentimental and
emotional ties fosters a spirit of competition, aggrandizement,
and mutual exploitation. . . . Frequent close physical contact,
coupled with great social distance, accentuates the reserve
of unattached individuals toward one another and, unless
compensated by other opportunities for response, gives rise to
loneliness. The necessary frequent movement of great
numbers of individuals in a congested habitat gives occasion
to friction and irritation. Nervous tensions which derive
from such personal frustrations are accentuated by the rapid
tempo and the complicated technology under which
life in dense areas must be lived (Wirth 1938:12, 15–16).*

In spite of insightful critiques, such views have persisted among
many people. Because of this stereotype, as well as the poverty and
suffering of slum dwellers in both industrialized and developing
areas, there is a tendency to deplore the rural exodus, the flight from
country to city, and to interpret it in pathological terms. Actually, as
we shall see, for all the squalor of urban shantytowns in the cities of
developing countries, most rural migrants feel that they are better off
than in their natal villages, most find jobs, and most have little or no
desire to return permanently to the country. For in addition to the
economic and other material advantages they find in cities, the faster
tempo of urban life, its excitement, and the greater freedom also attract them. Robert Park, although a member of the Chicago School
of urban sociology who shared many of Wirth's views, recognized
the innate attraction of the metropolis, which he found to be due, in
part,

*to the fact that in the long run every individual finds
somewhere among the varied manifestations of city life the
sort of environment in which he expands and feels at ease;*

*finds, in short, the moral climate in which his peculiar
nature obtains and stimulations that bring his innate
disposition to full and free expression (Park 1915:608).*

Park was thinking primarily of American cities, but his observations
apply equally to peasant emigrants to cities in the developing world.

### Causes of emigration

Why do villagers leave home to take their chances in cities? Students
of urbanization speak of the interaction of "push" and "pull" factors,
the combination of negative conditions in the country and the attractions of the city. The push appears due, above all, to rapid population increases in the country without corresponding development of
resources. With the conquest of smallpox, malaria, whooping cough,
and the other killing diseases that formerly held down population
growth, rural populations in most of the world are doubling every
25 years. The figures for Tzintzuntzan, Mexico, are not atypical for
much of the world: Between 1940 and 1960, the death rate dropped
from 30 per thousand to 9 per thousand, while the birth rate remained constant around 45 per thousand. In other words, the net increase per thousand jumped from 17 to 36, an annual population increase of 3.6 percent, one of the highest in the world. But during
these two decades in which population climbed from 1,000 to nearly
1,900, there was no increase in the already scarce supply of arable
land. Without the temporary safety valve of bracero contractual labor
in the United States, which drew off a hundred or more men annually over a period of 20 years, it is difficult to see how the village
could have survived. Now that the bracero program has ended, it is
only through massive emigration to Mexico City and smaller urban
centers that the surplus population can be cared for.

A lesser but by no means unimportant "push" in Tzintzuntzan
is manifest by those men who run afoul of the law, usually as a consequence of murder, and who, with their families, flee the village.
Although Tzintzuntzan is far from a violent village, I estimate that as
much as 10 percent of emigration during recent decades has been due
to the fear of vengeance by relatives of the victims.

The "pull" of cities is manifest first and foremost in the possibility of economic gain. Wages are much higher than in villages, and
in spite of the obvious unemployment and underemployment in the
cities of developing countries, it is astonishing how many migrants
find employment. In his Buenos Aires study Germani found that "The

most powerful motive for migration—the search for employment and better working conditions"—was rewarded with success. The majority of migrants "found work within a fortnight of their arrival, others a little later, but all or almost all did find work. Furthermore, the groups investigated consider working conditions at Buenos Aires decidedly superior to those in the areas from which they emigrated" (Germani 1961:228). Kemper found a comparable situation in Mexico City where, in spite of their lack of skills, nearly every Tzintzuntzeño who migrated to Mexico City and who genuinely was committed to staying found adequate employment (Kemper 1971:114). The pattern must be comparable in other great cities of the developing world— in Lima, Rio de Janeiro, Cairo, Bombay, Bangkok—for without employment these metropolises could not grow. The other "pull" attractions of cities are likewise fairly obvious: better schooling, better medical facilities (often achieved through social welfare plans that function in cities but not in rural areas), a higher standard of living (with electricity, propane gas, and so forth), and more interesting things to do. The importance of the latter cannot be overemphasized. However attractive the countryside may seem, especially to the sophisticated urbanite not faced with the problem of grubbing a living in a squalid village, the fact remains that the city is for most people far more interesting, and far more exciting, than is the village. Whether it is the Iowa farm boy or the Peruvian peasant, once exposed to the attractions of the city neither is anxious to return to his childhood home.

## The route to the city

Until fairly recently the fear of uprooting oneself from a small village and migrating directly to a huge metropolitan area seems to have inhibited a great deal of direct migration from village to city. Rather, people followed what sociologists have called a "two-step" migration pattern, in which the first generation of emigrants moved to a neighboring provincial town or market city, while their children then took the second step, to the metropolitan centers. In a widely cited study carried out in 1955 the Argentine sociologist Gino Germani studied migrants living in a shantytown on the outskirts of Buenos Aires. He found that most of the migrants were small-town rather than village people. Only 15 percent came from communities of less than 2,000, more than a third came from towns in the 2,000 to 20,000 range, and more than half came from larger communities (Germani 1961:212).

Matos Mar has noted a similar pattern in Peru, where until fairly recently railway and highway networks have been limited: The flow of population has been

> from the agricultural communities or rural areas to the nearest centres of population, from there to the largest town of the region and finally to Lima. The trend is generally from the rural mountain areas to important towns in the large valleys of the Andes, and from there to the coastal region (Matos Mar 1961:173).

There, in large towns and cities such as Piura, Chiclayo, Chimbote, Ica, and Tacna, *serranos* learn their first major lessons in urban life, after which many continue to the ultimate goal, Lima.

The two-step migration pattern in large degree is a function of a prehighway (and to a lesser extent, prerailway) era, for increasingly the data show that today's emigrants go directly from their point of origin to their ultimate destination rather than stopping for a decade or a generation at a way point. The importance of motor vehicles is indicated by Caldwell's exhaustive study of rural–urban migration in Ghana:

> Until the mid-1930's walking was an important means of migration, but in recent times it has been of little importance. Road transport, and above all mammy lorry transport, completely dominates the picture, although by 1963 competing bus routes were beginning to spread (Caldwell 1969:128).

Thanks to roads, emigrants can make trips of several hundred miles with less difficulty than their parents had in migrating for fifty or less. And with the recognition that real opportunity is far greater in large cities than in provincial capitals, it is not surprising that a majority of emigrants opt for the former destination. When a villager decides to leave home today, he has much to gain and little to lose by trying for the greater prize. In Bogatá, Columbia, a recent study of a sample of 200 migrants to two *barrios* illustrates the extent to which the two-step (or multistep) migration pattern is giving way to the single movement: More than two-thirds of those interviewed had come directly from their rural point of origin to the city; an additional eighth had made one stopover en route; and only a fifth had made two or more stopovers (Flinn 1971:83–85).

Another factor stimulating direct emigration is that the city, while still a foreign world, has become much less strange to villagers than it was a generation ago. Radio and especially television provide information about urban life previously lacking to peasants. This, and increasing opportunity to make short trips to cities to visit with relatives who have previously migrated, removes much of the traditional apprehension felt by country people for urban centers. The importance of previous city visits, and consequent familiarity with city life, as a factor in permanent migration is suggested by a recent study of 900 males in Monterrey, Mexico: Nearly two-thirds of the men had traveled to Monterrey at least once prior to their definitive move (Browning and Feindt 1971:54).

## Settling in cities

Regardless of how a migrant reaches the city, he is faced with the problem of finding a place to live, finding work, and finding a satisfying social life. In Mexico, and perhaps in most of Latin America, the majority of migrants seem first to stay with relatives while they find work and living quarters. In the Monterrey study just cited, 58 percent of newly arrived immigrants stayed with relatives, while an additional 11 percent lived independently but had relatives in the neighborhood. Only 31 percent lived independently without relatives in the neighborhood or in the city (Browning and Feindt 1971:64). In societies where a mat on the floor is a common bed, an influx of country cousins does not cause the hospitality problems that American middle-class city dwellers might encounter. It is taken for granted that a family member lucky enough to have living quarters in the city will take in relatives and assist them until they find lodging and work. Young people in some instances live with uncles and aunts for years while pursuing higher education.

In a comparable Ghanaian sample, only 27 percent of migrants stayed with relatives other than spouse, parents, or children, while another 27 percent stayed with friends, usually fellow villagers, a category not listed in the Mexican study. Another 20 percent initially rented accommodations (Caldwell 1969:130). Other studies suggest, but do not conclusively prove, that Latin American migrants are somewhat more likely to rely on kinship ties for settling in cities than are Africans.

Migrants usually find employment through informal family and friendship networks. In a well-known study in Mexico City, Butterworth cites the case of a Mixtec Indian from a Oaxacan village who worked up to be supervisor of general services in an unnamed company, in which role he found employment for 21 of the 31 fellow villagers who in subsequent years came to Mexico City (Butterworth 1962:261). Kemper also finds that kin and friendship networks are the single most common path by which an immigrant finds work (Kemper 1971:116).

## The role of voluntary associations

In many parts of the world so-called "voluntary associations" are major acculturative devices for rural peoples. These are perhaps most common in Africa where their multiple roles include help in finding employment and learning the ropes of city life, in social control, mutual aid, recreation, and capital accumulation, and in maintenance of village ties. Little has described the way in which African societies aid the raw tribesman:

> The newly arrived immigrant from the rural areas has been used to living and working as a member of a compact group of kinsmen and neighbors on a highly personal basis of relationship and mutuality. He knows of no other way of community living than this, and his natural reaction is to make a similar adjustment to urban conditions.
>
> This adjustment the association facilitates by substituting for the extended group of kinsmen a grouping based upon common interest which is capable of serving many of the same needs as the traditional family or lineage. In other words, the migrant's participation in some organization such as a tribal union or a dancing compin not only replaces much of what he has lost in terms of moral assurance in removing from his native village, but offers him companionship and an opportunity of sharing joys as well as sorrows with others in the same position as himself. . . . Such an association also substitutes for the extended family in providing counsel and protection, in terms of legal aid; and by placing him in the company of women members it also helps him to find a wife. It also substitutes for some of the economic support available at home by supplying him with sickness and funeral benefits, thereby enabling him to continue his most important kinship obligations. Further, it introduces him to a number of economically useful habits and practices such as punctuality and thrift, and it aids his social reorientation

*by inculcating new standards of dress, etiquette, and personal hygiene. Above all, by encouraging him to mix with persons outside his own lineage and sometimes tribe, the voluntary association helps him to adjust to the more cosmopolitan ethos of the city (Little 1957:592–593).*

Voluntary associations are common in Lima, where they tend to be based on a common geographical point of origin rather than the tribe, a concept far less meaningful in most of Latin America than in Africa. Nevertheless, association patterns are strikingly like those of Africa, and similar roles are fulfilled.

*One of the most important aspects of the clubs is the role they play in acculturating the serrano to life in Lima. . . . Many Lima customs are learned in the clubs and unacceptable customs, or at least those marking the person as rural, Indian, and serrano, are discouraged. The most visible traits, such as hair style, coca chewing and the more obvious clothing differences disappear first, and in most cases men change before women. . . . It is surprising how fast migrants learn the significance of things like hats and braids, also, how willing most of them are to discard them (Mangin 1959:28).*

In Mexico City, in contrast, comparable associations seem completely absent. Oscar Lewis found no clubs among the peoples he studied (personal communication), and Butterwork writes of the Mixtec migrants with whom he worked: "One of the most striking aspects of the group of Tilantongueños now living in Mexico City is the complete absence of any formal or informal participating in organizations" (Butterworth 1962:261). Tzintzuntzeños in Mexico city likewise belong to no association based on common origin, and very few belong to any club at all (Kemper 1971:155).

Lack of interest in Mexico in the kinds of voluntary associations so widespread in Africa may be due to basic differences in structural principles underlying social organization. In Africa so-called "corporate" kinship and tribal organizations are widespread, and the sense of association is inherent as well in the secret societies of West Africa. In Mexico, on the other hand, in both rural and urban areas social "networks," based on the concept of what I have called the *dyadic contract* (Foster 1967a:212–243), or personal formal and informal contractual understandings, underlie the structure of society. A corporate sense, manifest in voluntary associations, and informal interpersonal networks are both satisfactory devices for building lifetime relationships as well as learning to adjust to urban life; and the group that favors one can play down or entirely ignore the other, thereby avoiding duplication of function.

Mexico City dwellers, in spite of their relative lack of use of the associational principle, do participate in at least one type of association with counterparts in much of the rest of the world. The Mexican *tanda* (Davis 1968) conforms to the generic, capital- raising type described by Geertz as a "rotating credit association," found widely in Asia, Africa, and the Americas (Geertz 1962). In basic form, these associations are very similar. A group of members each week (or period) contribute the same amount of money to a common fund, drawing lots to see who receives the pot on each occasion, until all have received back a sum equal to their payments. In societies where Anglo-Saxon thrift is unknown and, in fact, usually unworkable, this system of enforced saving makes it possible for people to accumulate considerable sums of capital that might otherwise be beyond their means. Whatever their forms, it is clear that the voluntary association is, in most of the developing world, a major factor in facilitating the urbanization process.

### The quality of city life

Perhaps because of the influence of the Chicago School of urban sociology, American social scientists have tended to deplore the rapid urbanization of the developing world's countries and to point to the shantytowns that are a part of their large cities as evidence of social pathology. Many feel that major efforts should be made to hold rural peoples in the country by improving the quality of village life, particularly by providing schools, better health services, and other amenities. Paradoxically, such efforts, which I wholeheartedly support, probably have just the opposite effect to that intended since they further whet the taste of village peoples for the good things of urban life, and by better preparing young people to compete in the contemporary world they make it easier for them to find employment away from home. Rather to the surprise of many critics, urbanization studies indicate that, at least in Latin America, migrants are much better off economically in the city than in the country, much more satisfied with life in general, and better adapted to the realities of the modern world. The social disintegration, the loneliness, the anomie that Wirth and others assumed to be a necessary part of urban life, while by no means lacking, are far less common than one might suspect. The positive features of migrant life in cities, on the other hand, are much more marked.

In his Buenos Aires study Germani found it necessary to question the assumption that families in rural areas exhibit a greater degree of stability and attachment to traditional family values than do urban dwellers. Although some previously well-knit family units were found to have disintegrated in the city, the majority of families acquired urban ways and benefited positively from the experience. Whereas recently arrived families were marked by a strong authoritarian flavor, city-born and long-settled migrant families displayed "a relatively unconstrained atmosphere, a greater degree of friendliness among the adult members of the family, and more co-operative and democratic behavior" (Germani 1961:215, 218).

While not glossing over the grimmer aspects of *barriada* shantytown life in Lima, Mangin notes:

> There is a relatively high degree of integration and 'belongingness,' and considerable pride in achievement and satisfaction with home ownership. . . . [Barriada living] represents a definite improvement in terms of housing and general income, and Lima represents an improvement over the semifuedal life of the mountain Indian, Cholo, or lower-class mestizo. There is very little violence, prostitution, homosexuality, or gang behavior in barriadas. Petty thievery is endemic throughout Lima, but barriadas seem somewhat safer than most neighborhoods (Mangin 1960:913, 914).

The author adds, however, that the reason for this last fact may be that there is less to steal.

In Mexico City Butterworth found that Tilantongo migrant families enjoyed a higher standard of living than in the village, their diet was much improved, drinking declined drastically, a high degree of village-type face-to-face relationships was maintained, and family ties "remain as strong as they were in Tilantongo, and perhaps even stronger" (Butterworth 1962). Similarly Kemper found significant improvements in the quality of family life among Tzintzuntzeño migrants to Mexico City:

> [Husband-wife relationships] tend to be egalitarian and mutually supportive: most men assist and cooperate in

*domestic chores, and progressive, younger women are career-
as well as family-oriented. . . . nearly all families demonstrate
a low level of male authoritarianism and a high degree of
"democratic" conflict resolution between spouses (Kemper
1971:100).*

At the same time, father-child relationships seem much improved over the village:

*Fathers are more apt to be affectionate toward their
children, open and understanding in their counsel, eager to
labor for their children's betterment, and concerned with
being a friend rather than a symbol of ultimate authority—that
is, they tend to reject the very behavior which is
associated with rural machismo (Kemper 1971:101).*

### Urban-rural ties

After a villager moves to the city, he is faced with divided loyalties
and emotional tuggings. There are his family members and friends
who have helped him find his niche in the city, and his new friends
with whom he becomes acquainted at work and in his neighborhood. On the other hand, there are relatives and friends, greater in
number, who have been left behind, as well as the psychological
attachment to the village, to its baptism and wedding feasts, and
to its religious fiestas. How do migrants cope with these conflicting
pulls? Two principal patterns of adaptation are apparent: Either the
villager rather quickly becomes an urban dweller, sloughing off
most of his village ties and returning to the village only occasionally,
perhaps for the annual fiesta; or he goes through a long period
of moving back and forth, remaining in the city for months and
then returning for an equal period to the village. Many men leave
their wives and children in the village for years, returning periodic-
ally to visit them and bring money. A major purpose of many of
the voluntary associations in Peru and West Africa is to foster a
continued interest in the local community, perhaps even marshaling
resources to help with community development.

Although no comparative study of the two patterns has been
made, the reasons for the difference seem fairly clear. In tribal
Africa, and in any other place where the migrant may share in
an undivided interest in village land or resources, it is obviously

to his interest to keep alive this claim to village or tribal membership, at least to the point where the cost in time and money surpasses possible benefits. Membership in a tribal association concerned with village affairs is an important way continually to reassert this claim, as are repeated visits to the home village. Furthermore, in areas where communal resources are scanty or lacking, the migrant who owns land or otherwise controls local resources obviously is going to find it necessary to maintain close touch with his village. In Tzintzuntzan, for example, migrants who own appreciable amounts of land, or whose wives are skilled potters, are more apt to leave their families in the village, returning home once a month or so for a long weekend, than are young, landless men, who almost always bring their wives to the city as soon as they have established a toehold. To the extent that we can generalize, it seems safe to say that when the village is not seen as a refuge, as a source of economic aid in time of crisis, migrants tend to strengthen city ties at the expense of rural relationships.

It would be erroneous to assume that all rural families solve their social and economic problems by moving to crowded cities. Clearly this is not so; there are failures who return home, as well as others who cling to the edges of urban life without ever meeting success. But the evidence does indicate that the urban aspirations of a very great number of rural migrants are realistic, and that to a far greater extent than is often supposed these migrants find a better and more satisfying life than they knew in the village. It is not necessary to gloss over the enormous problems that face mushrooming cities in developing countries. But in attempting to understand all of the variants involved in nationwide development, and to plan for this development, the city must not be saddled with an image and reputation that is untrue and undeserved.

## Implications of urbanization for social change

After examining the processes of urbanization in the developing world, we turn to the *implications* of the phenomenon, with respect to both emigrants and their peers who remain in the village. Here we are struck by the fact that change does not take place in a quixotic and unpredictable manner, but rather that the events of cultural modification are patterned. The noteworthy point about urbanization is that the common process of movement from village to city produces results that are much the same in all parts of the world, entirely apart from historical relationships often cited to explain similar changes. In other words, the processes of change in urbanization are independent of specific cultural forms; they are functions of similar circumstances. The identification of these cross-cultural regularities—the similar patterns of change through time—is basic to developmental work, for they set outer limits within which programs can be carried out, and they permit a degree of prediction that is essential in all planning. These regularities are not experimentally verified; they are derived empirically from the comparison of large numbers of cases. Hence, they are only as good as the data from which they are drawn, and there are always exceptions. The examples that follow illustrate some of the common processes that are associated, directly and indirectly, with urbanization, processes that affect both emigrés and those who remain in rural areas.

1. *Cities Are Focal Points of Change.* Most social and economic change begins in cities, usually among the upper classes; it spreads downward to the traditionally inarticulate urban lower classes, and then outward to the countryside. This was true of preindustrial life, where a large part of the content of peasant culture was made up of elements that had slowly diffused, over generations, from city to village. It is much more true today in the developing world, where because of rapid urbanization the city is much less remote from the country than in earlier years. Migrants, especially those who maintain meaningful rural ties, are among the most important channels of communication between city and village, the carriers of urban culture to the hinterlands. For India, Crane outlines the mechanism of transmission as follows:

> Because the city is a diffusion center for Western ways and because life there is relatively free from the tight social controls of the rural villages, the volume of transient labor in the large cities has special significance. Sojourners from the village carry back home with them new ideas, new attitudes, and new skills. These innovations, allied with other circumstances making for change and dislocation of the traditional ways that have for long characterized village life in India, stimulate the slow diffusion of elements of a different way of life from the cities out into the countryside. Just as the economic tentacles of the city spread out to the hinterland, so too do the intellectual and cultural influences (Crane 1955:467).

Coleman recognizes the acculturative role of the city in Nigeria, where rural emigrants had contact with Europeans and educated Africans and were exposed to industrial goods and gadgets.

> *It was in the city that new tastes and wants were created,*
> *new values adopted, and emulative urges asserted. The*
> *acculturating influence of the city was in turn carried to*
> *rural homelands through kinship associations and tribal*
> *unions, or by the vivid oral reports of returning migrants.*
> *Urban dwellers became acutely aware of the wide gap between*
> *the higher standards of living and the greater amenities of*
> *the city—especially in the European quarters—and the*
> *poverty of their rural villages. They therefore consciously*
> *endeavored to take the enlightenment, modernity, and*
> *"civilization" of the urban centers to the villages*
> *(Coleman 1958:73).*

The process is the same when looked at from the village. The Indian anthropogist, S. C. Dube, finds that the presence of a nearby big city is an important factor in determining the degree of acculturation that will take place in rural areas.

> *The inspiration and lead for modification in the traditional*
> *ways definitely come from the urban areas, brought into the*
> *village community by semi-urbanized people or inspired*
> *by the example of urban relatives. . . . The rural communities*
> *clearly take the lead from the urban areas, although not*
> *without hesitation, misgivings, doubts and an initial*
> *resistance (Dube 1955:230–231).*

The practical implications of this theoretical generalization are clear: Technicians working in rural areas cannot ignore the nature of the relationships of small communities with larger centers. The nature and kinds of relationships that exist between the village and the city may well hold the key to successful innovation. Friedl has pointed out that in Greece there is a good deal of upward mobility, and peasant sons educated in the city enter medicine, law, and other high-prestige professions. These men and their families maintain contact with their native villages and visits back and forth are frequent. Many new ideas and attitudes and changes in style of life are brought to village families by city relatives (Friedl 1959). Village sons who have acquired the knowledge and sophistication of cities might very well prove to be among the most effective allies of developmental workers in rural areas.

2. *Growth of a Worldwide Monetary Economy, Stimulated by Urbanization, is Producing Major Changes in Family Structure.* Among preindustrial peoples some kind of large or "extended" family is often the functional unit of social and economic interaction. Obligations and expectations of reciprocity extend to distant uncles, cousins, and nephews and nieces, and social gatherings may embrace scores of related people. Frequently the social, psychological, and economic security of the individual stems from membership in an extended kin grouping. The individual is tied to his family, but it is also his source of strength.[1]

Today, however, the growth of a worldwide monetary economy is drastically changing these patterns. Cash crops for sale in cities are replacing subsistence crops, and increasing opportunities for wage labor on plantations or in factories are the rule. Greater independence is thus possible for those who wish to break away from family ties. The mutual security functions of the extended family are less important than formerly—at times, as we shall see, they are a positive handicap—and so the functional family group shrinks in the direction of the Western nuclear biological family of parents and children.

This trend is reflected in preindustrial societies of all degrees of complexity. Hamamsy finds that in the American Southwest among the Navaho Indians at the Fruitland Irrigation Project wage labor is increasing the importance of the independent biological family, whereas the influence of the extended family and of larger tribal units lessens (Hamamsy 1957:105). In nearby Cochiti pueblo Lange discovered that clan organization is breaking down in functional importance; looking forward he predicts that it will disappear from general Cochiti consciousness, to be replaced with kinship concepts and terminology from Spanish and Anglo cultures (Lange 1953:693).

At the level of the peasant village the process is the same. In Kharga Oasis in Egypt, Abou-Zeid found that as a consequence of widespread migration to the Nile Valley for wage labor, family bonds "suffer badly and have become less effective in everyday life" (Abou-Zeid 1963:50). The size of the family has diminished considerably and the nuclear type has become more frequent, "even

---

[1]For reasons that cannot be discussed here the importance of the extended family in southern European and Latin American rural communities is somewhat less than in many other parts of the world.

dominating the scene in Kharga town" (Abou-Zeid 1963:50). The picture is the same in Nigeria where Coleman finds that the shift from a subsistence to a money economy has "loosened or extinguished the coöperative ties binding the individual to his clan or lineage members. Lineage bonds have been corroded by the strong attractions of the individual profit motive and by extralineage opportunities" (Coleman 1958:67–68).

At a more advanced stage of change—in Japan, for example—the same trend continues. There, as in other nations that have undergone industrialization, ties of kinship have diminished rapidly.

> In the period of less than a century since the opening of
> Japan to the West, changes in kinship have followed a
> familiar pattern. The nuclear family is displacing the extended
> family and has come to be the dominant form in the cities
> of Japan. Where the extended family exists, it has everywhere
> diminished in size and functional importance.
> (Norbeck and Befu 1958:116).

These worldwide changes in traditional kinship patterns do not represent a simple case of the diffusion of Western forms. Rather, we are dealing with a cause-and-effect situation. The demands of urbanization, industrialization, and a monetary economy usually are uncongenial to the extended family groupings of subsistence peoples. Where these changes come about, we can predict that traditional family forms almost certainly will be modified in the direction of the nuclear biological family. Gough is describing a worldwide phenomenon when in speaking of the Nayars of Malabar, India, she writes:

> Kinship change among the Nayars is not explicable
> in terms of the concepts of "culture contact" or "cultural
> borrowing," but rather in terms of growth in the social
> structure as a whole, stimulated by external economic factors.
> Changes in the Nayar kinship system, correlated with changes
> in local organization, appear to have taken place in response
> to changes in the technology and economic organization
> of the society as a whole. They can therefore only indirectly
> be attributed to European contact. . . . It appears that the
> stage of disintegration of the traditional lineage system
> and of development of the modern bilateral system depends
> on the degree of absorption of the inhabitants into the modern

*economy of cash crops, cash wages and urban occupations, and on the consequent degree of social and spatial mobility (Gough 1952:86).*

In Chapter 6 we shall see that the traditional obligations inherent in large family groupings are often strained to the breaking point by this process of transition. At a particular point in the sequence, before a final break, they constitute a serious deterrent to the introduction of new and improved production techniques, since a progressive individual's greater income may be drained away in maintaining traditional forms of hospitality and help to relatives. The technician who is familiar with traditional family forms, and who has some idea about how far changes have gone, will be well prepared to decide whether his program is going to encounter difficulties because of conflicts in family organization. The technician who is ignorant of such factors will move blindly.

3. *An Urban-Stimulated Market Orientation for Cash Crops Is Destroying Traditional Rural Cooperative Work Patterns.* Anthropologists have noted cooperative work groups in village societies in many parts of the world. In West Africa and in Haiti large numbers of field hands may often be seen working to the accompaniment of music and drums in reciprocal activities known as the *dokpwe* (in Dahomey) and the *combite* (in Haiti). Other less colorful, but nonetheless important, reciprocal forms have been found in most of Latin America and in much of Asia as well. Analysis has revealed that some of these groups have significant economic functions and that total productivity is increased as a result of their role. But perhaps more often the social functions outweigh the productive functions, and in a "rational" sense the activity is uneconomic.

It is important to remember that most cooperative activities of this type do not represent work for the general welfare of the community. Rather, they are the expression of carefully worked out reciprocal relationships between individuals and families, in which obligations and expectations return, over time, to each participant about what he has put into the effort. Villagers have a strong *quid pro quo* sense. When cooperation of this type produces an equal amount of help for every man, plus social satisfactions, it is obviously attractive. These conditions tend to prevail in traditional communities. But when some farmers begin to forge ahead of their fellow villagers, the system is thrown out of balance, for the social output may come to outweigh the work input.

Erasmus, surveying the evidence from Latin America, found a high correlation between the commercialization of agriculture and the use and availability of money, on the one hand, and the decline in the frequency and intensity of reciprocal agricultural work patterns on the other hand (Erasmus 1955). With the attraction of a market, a peasant farmer usually becomes quite "rational." When he finds it is cheaper to hire a few hard-working peons instead of paying for food and drink for a larger number of fiesta-minded friends, he is ready to let this aspect of tradition slip into the past. Some cooperative forms do continue, but usually because of the social values that still exist, only rarely because of economic values.

An understanding of the relative importance of cooperative work forms is important to developmental workers because of the frequent, but often erroneous, assumption that village people are "naturally" cooperative, and that if they prove reluctant to give the outsider the degree of group support he expects, something must be wrong with his presentation of the program. Chapter 2 gives reasons why villagers often are not basically cooperative, and why such degrees of reciprocal services as have been evidenced in the past are, in most cases, rapidly disappearing. When the technician is familiar with traditional cooperative work forms and knows to what extent they still exist and what social functions they serve, he will be less likely to jump to the easy conclusion that his new friends obviously like to work together.

4. *Migration from Rural to Urban Areas, with an Accompanying Shift from a Subsistence to a Monetary Economy, Frequently Produces Dietary Deterioration.* Through long experience, primitive and other subsistence-based peoples usually have learned to exploit their environment to obtain a relatively balanced diet. In addition to staples, they eat many things that Europeans and Americans would not consider to be food: wild seeds, fruits and berries, herbs and leaves, and insects. Illustrative of the latter are *jumiles,* flattish hexagonal Coleoptera that are a popular early spring delicacy among the lower classes in central Mexico. Lightly toasted, sprinkled with lime juice, and eaten with tortillas, or ground raw into hot sauce, these "flying bedbugs" are believed to strengthen anemic children and to make adults feel healthier. Fed to hens, they are reputed to produce more eggs with thicker shells. *Jumiles* are, it appears, rich in protein, and perhaps calcium as well, excellent additions to any diet.

Some years ago nutritionists became interested in rural Mexican diets. By American standards there appeared to be serious defi-ciencies: little citrus fruit for vitamin C and no milk for calcium. But investigation proved that the lime water in which maize was soaked before being ground and made into tortillas provides a good substitute for milk calcium, and vitamin C exists in abundance in the hot chile peppers eaten by many Indians. Small amounts of meat cooked with beans releases vegetable protein, providing sufficient protein with less meat than required by American standards. Wild herbs, grasses, and other items often added only for seasoning also contribute to the balance. In short, rural Mexican diets proved to be surprisingly good.

Rural peoples who grow and collect most or all of their food are not aware that they have a balanced diet, or even that there is such a thing. They do not realize that they are making daily decisions that have an important bearing on their health. When, however, they move to cities where the major part of their diet must be purchased, tradition and experience serve them less well. They must learn anew to develop a balanced diet, this time by allocating their food budget wisely. This is not easy, and it requires education and understanding. Only in recent years has real pro-gress been made in improving food habits in the United States, despite almost universal literacy, abundant food supplies, and formal nutrition education. Previously subsistence people, when faced with a variety of new foods, are, in a sense, starting out to invent a dietary culture. Guides are few, and the factors of taste and prestige tend to be dominant in determining new habits. Con-sumption of sugars and other items with attractive sensory qualities increases rapidly. Processed and packaged foods, because of the prestige attached to them, become more important than formerly, even though in the processing much of the natural nutritional value has been removed. Desired and desirable foods such as meat, eggs, butter, and fruits often are obtainable in only small quantities because of high costs. Since malnutrition develops slowly, peoples in the process of learning new dietary patterns usually are not aware of the relation between their health and their food.

The frequent contrast between well-nourished rural peoples and less well nourished urban peoples is illustrated by the African work of the pediatrician, D. B. Jelliffee, and his colleagues. In one study in northern Tanzania all children in the small tribe of Hadza hunters were examined. It was found that "the clinical nutritional status of all the children was good by tropical standards; in particular, the syndromes of kwashiorkor and nutritional marasmus, rickets, infantile scurvey, and vitamin B deficiency syndromes were not seen" (Jelliffe et al. 1962:911). A parallel study of several hundred Acholi children, members of a large Nilotic group in Uganda including both village and city dwellers, revealed that the rural children had significantly less protein-calorie malnutrition than did the urban children. The authors believe this difference stems from the substitution in cities of maize flour and other predominantly carbohydrate foods for the traditional vegetable protein foods such as millet, sesame, and beans, not easily acquired in towns (Jelliffe et al. 1963).

Speaking of the general problem in Africa, and referring specifically to infant and childhood nutrition, Jelliffe and Bennett summarize the picture:

> Failure of lactation appears as a result of imitation of presumed social superiors, ill-advised advertising of dried milk, and of mothers going to work in employment where a suckling baby is not aesthetically acceptable. As a consequence, in tropical towns the growing double scourge of infective diarrhoea and nutritional marasmas is increasing rapidly, as a result of attempted bottle-feeding with milk formulas of homeopathic strength, but heavily bacterially contaminated. In addition, the foods available to the "new townsmen" immigrants are usually far different from those growing in the home villages, and are often more limited in range and of lesser intrinsic value (e.g., cassava or maize flour instead of millet); while, at the same time, the often quite wide variety of semi-wild foods, including certain green vegetables, insects, etc., will no longer be available (Jelliffe and Bennett 1963:16).

Knowledge of this common—but by no means universal—pattern of dietary deterioration in developing countries can be of much help to public health workers. It should be clear that in areas still producing a high percentage of the food consumed, and where industrial products are not yet easily available, environmental sanitation and basic immunization efforts should take precedence over nutrition education. But where cash crops have become important, and where wage labor in cities, or mines, or in the country is the rule, concern with diet should mark any health education program.

## NATIONALISM

The second of the major forces of modernization to be considered in this chapter is nationalism. Without an understanding of the nature and processes of nationalism, especially in its cultural—as contrasted with its political—dimensions, we cannot comprehend fully the impact of the modern world on developing nations and their cities and their villages. Nationalism, as an historical and political phenomenon, is relatively recent, associated especially with the name of the German philosophor and poet, Johann Gottfried Herder (1744–1803). Historical nationalism first appeared as a major force in Western Europe in the late eighteenth and early nineteenth centuries, when small principalities and dukedoms began to coalesce into modern nations. New nations, asking the question as to who and what they were, made use of the concept of a folk soul or folk spirit, a *Volkgeist,* the thread of continuity by which the present was linked to an ancient past. Evidence of the uniqueness of a nation's folk spirit was found in such things as popular language (as contrasted with French), in folklore, folk song, and folk traditions. In these grassroots traditions, newly nationalistic peoples found evidence of a creative national spirit, a vitality peculiarly their own, not shared with other peoples.

In all instances the need was to emphasize the antiquity of the forms, to prove that historical and cultural roots reached far into the past. Greek independence in the early nineteenth century was fostered by concern with and pride in former greatness, and by the visible evidence of past high civilization. Not all European nations were similarly endowed with reminders of classical antiquity, but through the search for folk elements a functional substitute was found. Not surprisingly, ethnographic research proved to be an important tool in ferreting out evidence of cultural antiquity, and it is to this need that we owe the origin of much ethnography. Hofer, pointing out that systematic ethnographic studies began in the early nineteenth century, writes:

> The centres of ethnographic growth were those regions where the creation of national states and cultures had become a vital problem (for example, Germany). Herder and the Grimm brothers are usually credited as initiators of the new discipline. Generally, Central Europe is considered to be the birthplace of ethnography (Hofer 1968:312).

Hofer, who correctly sees nationalism as a revitalization movement, has described the process in his native Hungary, where the reform era began during the first half of the nineteenth century:

> *The reform of the literary language was launched. Attempts were also made to reform economic life, the civil service, law, art, and literature. The poets introduced national metrical structures extracted from folklore, and wrote epic poems of the past; these replaced older epics which were not felt to fit in with the political movements for national independence and social reform. The sources of the new national culture were sought in national history and folk culture or "small traditions"* (Hofer 1968:312).

Although nationalism usually is studied as a political and historical phenomenon, to the anthropologist it is simply a variant—albeit an enormously important one—of the acculturation process. In acculturation situations, ranging from primitive tribes in fresh contact with more advanced peoples to developing countries with a flood of new relations with the industrial world, there seems to be a sequence through time in the form of a series of stages of contact. This sequence was outlined by Elkin in 1936 (Elkin 1936–1937). More recently Read, speaking of European education in Africa, has described the same phenomena (Read 1955:105–110). In the initial period of contact the recipient people may be antagonistic to changes suggested or imposed by outsiders, other than in the field of material goods of obvious utility. Adults are fearful of the changes they realize will come.

Subsequently there is increasing acceptance of the outsiders' ways, particularly by the younger generation, and growing enthusiasm to learn more. This leads to the rejection of a great deal of indigenous culture, to scorn for traditional ways, and to disregard of the advice of elders. The validity of earlier customs is denied, and people who cling to old ways are taunted as old-fashioned. There is a headlong rush to acquire foreign culture, and a desire on the part of the local elite to become like the economically dominant outsider.

In much of Africa, traditional songs, dances, and folklore were put aside, and hymns, drill, and school readers took their place. As a paramount chief told Read, "The white teachers taught us to despise our past" (Read 1955:108). For a time educated Africans did just that.

This initial period is followed by a period of disillusionment. It soon becomes apparent that the members of the less complex society cannot participate fully in the society of the more complex group. Restrictions imposed by the dominant power are part of the cause, but other deeper cultural and psychological causes are also involved. The dominated group feels that its own culture is threatened, but it has nothing to substitute; feelings of insecurity result. In some regions, in parts of Oceania, for example, depopulation followed; but more often, among tribal peoples the common reaction is the nationalistic manifestation of nativism. Several forms may occur, but the common element is a partial or complete rejection of the culture of the foreigner and an attempt to return to or restore the fundamental values of earlier days. The Ghost Dance of 1890 of the North American Indians is one classic example of a nativistic movement. The Cargo Cult reactions of Melanesia in the first half of the twentieth century is another.

The same general sequence is apparent at the level of modern states. Newly developing countries, which, in the twentieth century, are anxious to assimilate the material techniques of the West but, at the same time, to maintain indigenous spiritual values, find themselves frustrated in that changes do not come as easily as they wish. The elite, who a few years earlier eagerly sought to identify themselves with the ways of the West, in dress, education, food and politics, and who often deprecated their own culture and its achievements, now lead their people in a search to discover the essence of their traditional cultural forms. The values inherent in ancient ways are recognized, and attempts are made to restore and perpetuate them. At this point nationalism is well under way.

Ideologically, a nationalistic movement represents a group of people in search of its ethos, its unique essence, its being. Nationalism is a response to the disillusionment of people when they realize that the ways and values of outsiders will not fully serve them. It is then that they turn inward and ask, "Who are we? Whence have we come? What is our culture?" Nationalism represents a cultural position to which the members of the group are committed, and which serves as rationale and validation for their views and actions. Simmons, in describing the lower-class creole (mestizo) culture of coastal Peru, has given us a very good description of the cultural outlook to which the members of this nationalistic group are oriented. "The essence of this outlook," he says, "is its explicit affirmation of the uniqueness and originality of the [Peruvian] mestizo culture" (Simmons 1955a:109). Peruvian writers who attempt to explain *criollismo* speak of it as the creole "spirit," "way of life," or "soul," something that cannot be concretely defined, but "which stamps mestizo culture and mestizo personality within a particular identity and integrity of their own. *Criollismo* is the mestizo's answer to the painful question of who and what he is, his assertion that his 'way of life' is a positive creation of his own rather than a casual European-indigenous mixture" (Simmons 1955a:109).

Successful nationalism means the identification of a way of life as peculiarly one's own, a positive creation of the people concerned, and not as an importation or borrowing from others. This identification is achieved by manipulating a series of symbols, which must share two basic qualities: (1) a high degree of visibility, and (2) real or imagined antiquity, stemming from the traditional culture. These symbols are the focal points around which people rally, both to be convinced of, and to reaffirm their faith in, the vitality and uniqueness of their own culture. The symbols of nationalism that reappear time after time, in Latin America, Africa, India, and Southeast Asia, are surprisingly similar: language, costume and grooming, dietary patterns, fiesta celebrations, an interest in archaeology (which gives the best possible evidence of past greatness) and folklore (to reconstruct music, dance, and popular arts), humor, and sometimes folk medicine, sports, and religion. The surprising uniformity of symbols in all parts of the world seems based on limited possibilities rather than diffusion: There are only so many categories of behavior that combine the necessary qualities, and most of them are hit upon by all nationalizing peoples. Several of these symbols will be used to illustrate the ways in which nationalistic countries exploit them.

1. *Language.* Nothing can be more uniquely a society's own than a means of verbal and written communication not shared with or drawn from other societies. Not only is language a symbol of an ethos, but, equally important, it is the principal vehicle by which this ethos is built, spread, and perpetuated. It is for this reason that languages such as Tagalog in the Philippines and Hindi in India are so important as carriers of nationalism, and why Hebrew and Gaelic have been re-created in Israel and Ireland. With the first major stirrings of nationalism, the search for the appropriate language becomes of primary importance; in Europe, in response to these forces, first Latin and then French were dethroned as the speech and written language of the educated classes, in favor of local languages. Sometimes the selection of a national language

is based more on emotion than on logic, and wherever a world lingua franca, such as English or French, loses out as a consequence, problems of international communication—and often communication within a multilingual society—become greater. Nevertheless, when a vigorous local language exists, even though spoken by a minority of the inhabitants of a country, the odds are very good that it will be raised to the status of a national tongue. We have seen this happen with Hindi and Tagalog, and in Africa it is far from certain that English and French will maintain their preeminence among educated people.

It is interesting to compare Africa with Latin America: In preconquest times both areas had myriad local languages and cultures; both were conquered and ruled by European countries; and in both areas most nations ultimately became independent. But three centuries of colonial rule in Latin America firmly implanted Spanish or Portuguese speech, and in most countries there is no serious thought of returning to a native language. As a consequence, these countries have easy communication among themselves and with a large part of the world, and they are spared the linguistic uncertainties that characterize many newer lands. In contrast, European domination in Africa lasted a much shorter time, and, although widespread, English, French, and German never achieved the near-universal use of Spanish and Portuguese in America. Will the new African nations be able, or willing, to hold on to French and English? Or, under a powerful urge to have their own national languages, will they deemphasize the foreign tongues to favor local speeches?

2. *Dietary Patterns*. The emotional attitudes that can be expressed in food habits can scarcely be overemphasized. In every society the role of food in maintaining the solidarity of the group, and in affirming and reaffirming affective ties, is well marked. To offer food is to offer love, friendship, affection; to accept food is to reciprocate the proffered feelings. It is not surprising, therefore, that food and drink, and the modes of taking them, invariably become one of the basic symbols around which a national ethos is built. As with other symbols, dietary symbols are, to the extent possible, based on ingredients indigenous to the country concerned. Even in the United States, which no longer thinks of itself as intensely nationalistic, food symbolism is apparent on our most unique holiday, Thanksgiving. Our ancestors, the Pilgrims, ate those foods that today are symbolically important, and they were able to eat them because they were indigenous to the land: the turkey, which

roamed wild; cranberries, unknown in Europe; pumpkins, raised by the Indians; and corn, likewise cultivated by the Indians.

Consciously and unconsciously nearly every nation searches out its unique or ancient foods and endows them with symbolic status. In Mexico the symbolic food par excellence is turkey in mole sauce, usually served with rice, beans, tortillas, and occasionally guacamole (mashed avocado) and tamales. Of all these dishes, only rice is foreign; all the rest are indigenous to Mexico, many first domesticated or cultivated there. The mole sauce is especially interesting, for its ingredients include vanilla and chocolate (both native to the Gulf Coast), and peanuts, tomatoes, and chile peppers, all widely cultivated in pre-Conquest Mexico. In Arab countries symbolic meals are based on lamb, in West Africa the native palm wine has acquired nationalistic values, and in Peru hot peppers, native maize, and local fish pickled in lemon juice are all seen as peculiarly national.

Not only are the foods themselves important, but the mode of eating has symbolic value. Since the knife and fork are Western implements, they may on occasion be rejected when a banquet has conscious nationalistic significance. In Mexico dexterity with the tortilla is admired, as of course is the case with chopsticks in the Orient. Arabs derive satisfaction from dipping into a common bowl, and among the Amhara of Ethiopia commensals together dip out of the *injara* (bread) basket, scoffing at what they consider to be "the excessively individualistic Western custom of sitting each person down to a separate place at the table and thus depriving the meal hour of what they feel should be its basically communal tone" (Levine 1965:246).

In probably all societies food is conceptually related to health and strength. Everywhere one finds a sense of history, traditional foods—the simple fare of the ancestors—are popularly believed to have special nutritional values. This is amusingly illustrated by a Costa Rican novelist:

> Our zeal to imitate the foreigner has made us abandon
> our good food. Badly nourished, we lack vigor and
> enthusiasm for politics, for sane polemics, enterprises and
> love. We lack today great writers, daring entrepreneurs.
> We won the war of 1856 because our soldiers ate totopostes,
> cheese, meat, dulce, coffee, and such solid nutrients.
> There is an intimate relation between love of country and
> a taste for picadillos and tortillas. . . . Do our women want

*the vote? Magnificent! But let us deny it to her who does not
know how to prepare a picadillo, tamales, beans, coffee,
and bake a chicken. We have lost our classic, suave,
nutritious, and savory national cooking (Biesanz and Biesanz
1944:35, quoting Chacón Trejos, Tradiciones costarricenses).*

3. *Costume.* A brief visit to the United Nations in New York
provides convincing evidence of the symbolic importance of clothing to nationalism: kenti cloth from West Africa, the barong male
shirt from the Philippines, the jinnah cap from Pakistan, the Nehru
jacket, Indian homespuns from the time of Ghandi. Although
clothing lacks the universal and deep-seated emotional impact of
food, because of its visibility and national variety it is an obvious
device to express cultural uniqueness. So important, in fact, is
clothing that in 1967 newly independent Zambia appointed a
National Dress Committee to advise the government on what should
be accepted as correct dress on formal occasions *(San Francisco
Chronicle,* November 3, 1967).

4. *Fiestas, Folklore, Dancing, Art.* Popular arts are of enormous
symbolic value in all new countries. Dancing, singing, and drumming
are available to all peoples, mass expressions of folk skills and folk
spirits. Because of this, folklore societies often appear with the first
stirrings of nationalism, and national ballets for local and international performance have come to have such an important function.
In Peru one notes the highly stylized song and dance form, the
*marinera,* and the *vals criollo,* the recast European waltz; in Mexico,
the *mariachi* orchestra and the *huapango;* in Spain, the flamenco
dancer and the *cante hondo,* and in Scotland, the bagpipe. Rare indeed is the country that cannot express its ethos in music and dance.
Plastic and graphic art forms, too, have enormous importance, and
much of Africa is especially favored in this sense, since the intrinsic
merit of African art was recognized by the West well before the development of nationalism.

5. *Archaeology.* Archaeology has proven to be an especially
powerful instrument in developing nationalistic sentiments in newly
independent or newly developing countries. It is a device by which
a nation may know its past, and it is particularly useful for nations
with short periods of written history. Grahame Clark points out the
relation between nationalism and archaeology when he speaks of the
latter's ability to foster sentiments needful to the stability and existence of society. Archaeology, he says, "multiplies and strengthens

the links which bind us to the past, and provides innumerable material symbols of social development through the ages, symbols all the more effective because visible and tangible" (G. Clark 1947:191). It is not for love of science alone, therefore, that national museums of archaeology, anthropology, and history are so high on the list of priorities of many countries. In spite of their costs they are wise investments in fostering a sense of patriotism and in calling the attention of foreign tourists to the cultural antiquity and richness of the countries concerned.

6. *Humor and Lively Wit.* Humor, and its associated personality characteristics, often constitute a nationalistic symbol, stemming as they do from the belief that foreigners can never fully appreciate the subtlety of wit that characterizes the ingroup. Members of a society often tend to stigmatize foreigners as more slow-witted, perplexed by the turn of phrase or thrust of wit that is obvious to their peers. Simmons puts this well in describing the Peruvian creole's self-image; he sees himself as a personality type marked by "a quick and brilliant mentality, a facile creative talent, and a profound sense of humor . . . consumate skill in repartee" (Simmons 1955a:110). In the contemporary world, movie and television stars often epitomize the self-image of quick wit and humor that people like to propagate: Cantinflas in Mexico, Fernandel in France, Charlie Chaplin in the United States, and Alec Guiness in England. Even comic books portray the image, as evidenced by the enormously successful *Supermacho* in Mexico.

7. *Traditional and Popular Medicine.* In countries with ancient and documented medical systems that stand in contrast to contemporary Western medicine, there is frequently an urge to raise the indigenous system to "separate but equal" status. This is particularly true in India and China, and in neighboring countries such as Nepal and Indochina. Opler quotes the report of an Indian government committee appointed to consider measures to increase the usefulness of the indigenous systems of the country, particularly Hindu Ayurvedic medicine, and Moslem Unani Tibbi. The report describes the *Tridosha* or "Three Disorders" theory of Ayurvedic medicine, *avta* (wind or flatulence), *pitta* (bile or gall), and *kapha* (phlegm or mucus), which when kept in balance promote health. The report continues:

This theory is the foundation of Ayurvedic science. Its embryological, anatomical, physiological, pathological and

*therapeutical conceptions rest upon this foundation.*
It has been the product of the genius of this country and has
not been borrowed from outside. *There are references to it
in the Vedic literature also. The humoral theory of the
Greeks was, perhaps, a bad adaptation of the* Tridosha *theory.
Evidently it was the fruit of numerous observations and
prolonged discussions spread over centuries. . . . By the time
these Samhitas* [ancient medico-religious texts] *of Ayurvedic
literature were composed before the beginning of the
Christian era, the* Tridosha *theory was firmly established
(Opler 1963:32); my emphasis).*

Here we have all of the necessary ingredients for a powerful nationalistic symbol: It is an indigenous system, not influenced by outside
sources; is of great antiquity, antedating Christianity; is at least semi-
scientific; and perhaps was copied in bungling fashion by an ancient
Western civilization. The governments of both India and Nepal today
maintain Ayurvedic schools and laboratories.

Two recent works on Chinese medicine intentionally and unin-
tentionally portray the significance of traditional medicine (including
acupuncture) in a nationalistic setting. Stressing use and knowledge
antedating Western equivalents, Huard and Wong discuss early scientific achievements. They describe silver paste used to fill dental cav-
ities: "This technique, known in China for some thirteen hundred
years, *did not become common in Europe* before Regnart (1818) and
Taveau (1837)." Again, Shen Jua (1031–1095), a doctor, architect,
agronomist and historiographer, "surmised, *before Kepler,* that the
apparent course of the sun round the earth was elliptical and not
circular. And finally, "To the Ch'ing dynasty [1644–1911] we owe the
publication of great encyclopedias, *four times the size of the En-
cyclopedia Britannica* and containing medical sections" (Huard and
Wong 1968:33, 40, 58; my italics).

The second work, by Croizier, places traditional medicine in the
context of historical Chinese resistance to the West, the sudden
crumbling of much of this resistance toward the end of the last century, and its resurgence under communism:

*Modern nationalists have seen Chinese medicine as proof of
China's creative ability in at least this one scientific
area. With science generally hailed as the key to national
survival, and almost all that science of foreign origin, claims*

*for the scientific validity of the native medical tradition have had a powerful appeal to cultural nationalism (Crozier 1968:6).*

[Moreover,]

*in the general attack on traditional culture, Chinese medicine suddenly became a very important part of that cultural tradition. To defend it became part of the defense of Chinese culture . . . The tradesman's art of traditional China became a national treasure to latter-day traditionalists, an integral part of the "national essence." (Croizier 1968:82).*

Renewed reverence for the ancient system came well before communism, and in 1930 a visiting League of Nations expert found 15 schools for traditional medicine. The strongest force for rehabilitation of traditional medicine came, however, with communism:

*The application of such phrases as "national cultural legacy" or "medical heritage of the motherland" . . . expressed the new value given it. Literature on Chinese medicine, both scholarly and popular, rapidly expanded (Croizier 1968:175).*

Other frequent symbols of nationalism include religion, sports, and perhaps mass calisthenics. Symbols of nationalism are important because pride in one's culture and belief in a nation's ability to progress are essential in developing strong states. A recent and enlightening example of how the leaders of a nation may build on nationalistic feelings comes to us from Peru. In opening the meetings of the International Congress of Americanists in Lima in 1970 General Alfredo A. R. Risueño, Minister of Education, drew upon the past to validate his government's plans, at the same time rejecting outside influences:

*Today in Peru we look backward towards our autochthonous cultural tradition, [which is] without relationship or similarity to others, which had the vigor to develop its values the length and breadth of an immense land, and without losing sight of its distinctive characteristics we [now] search for the most fruitful guidelines which from the past project themselves into the future. . . . In contrast to the multiplicity of popular and national cultures, and ethnic groups [in Peru], there took form in our urban centers an elite, foreign-influenced*

> *subculture, divorced from our indigenous cultural forms, the*
> *possession of a privileged classist minority which appropriated*
> *for itself the label "culture." The creation and development*
> *of this subculture,* without roots in the native society,
> *converted it quickly into a reflection of the traditions and*
> *cultural norms coming from the foreign, dominating*
> *metropolises, simply an appendix to the imported cultures*
> *(my translation; my emphasis).*

The nationalistic appeal can also be used in more specific ways. By casting contemporary plans in the mold of traditional civilization, leaders may innovate more easily than if they had to "sell" their programs as something entirely new. When in the mid-1950s India abandoned the rupee-anna-pice monetary system in favor of decimals, there was a good deal of opposition, even in Congress. One of the most effective arguments ultimately leading to success was that, millenia ago, Indians had invented the decimal system, so that the new proposal was, in fact, merely the return to a truly Indian form (personal communication from Harry Gupta). Similarly, Hodgkin has pointed out how, during the independence struggles in Africa, the nationalist leaders stressed that the parlimentary democratic institutions for which they were pressing "amounted in fact to nothing more than an adaptation to new conditions of the traditional, and essentially democratic, forms of political system which were to be found in most parts of pre-European Africa" (Hodgkin 1956:171).

Unfortunately, symbols of nationalism can be carried to excessive lengths, as when small countries insist on the costly prestige of an international airline, uneconomic steel mills, or elaborate laboratory and hospital equipment, which lie idle for lack of scientists and technicians trained to use them. More amusing than costly is the news release of the New Delhi-based "Institute for Rewriting Indian History," which argues that the Rigveda dates back 500,000 years! (*The Times* of India, August 2, 1965). Even in the United States similar exaggerated statements are made, sometimes for less noble purposes. When the Wine Institute held a press conference to kick off an advertising campaign, a nationalistic theme was used:

> *Wine is in the American tradition. A historical study will*
> *point out that the Mayflower that brought the Pilgrims to our*
> *shores was also used in the wine trade, that George*
> *Washington had a wine cellar, and Abraham Lincoln once*
> *operated a tavern (San Francisco Chronicle, October 9, 1969).*

On balance, however, nationalism is a powerful force for bringing change and progress, and the technical specialist who is aware of the nature of the phenomenon, and who knows his host country and the local symbols of primary importance, can both avoid mistakes and capitalize on opportunities that would otherwise be missed. Such knowledge will explain, for example, why the people of a village may be more interested in a new mosque or church than in a community reading room. If the symbolic needs of religion are adequately satisfied, then perhaps people will be ready to move on to other things that are, by definition, not parts of the traditional culture. Or, to cite another example, a specialist familiar with local nationalistic symbols may see ways to arouse enthusiasm for a community project by organizing it around a traditional fiesta, or by using puppets who, in language, costume, and wit, epitomize the group's self-image. When a specialist knows what people are proud of and understands why, he sees innumerable ways for using this knowledge to help them through the process of change.

# 4

## the dynamics of change: culture, society, psychology, and economics

All societies are constantly in a state of relative tension. Each society can be thought of as a host to two kinds of forces: those that seek to promote change and those that strive to maintain the status quo. These forces are locked in perpetual combat, the former trying to throw the latter off balance in order to gain the ascendency, and the latter trying to prevent this from happening. Since a tendency to change is fundamental in culture, it is obvious that in the long run the forces that promote innovation will have the edge over those that strive for conservatism. But the degree of ascendency that change-making forces achieve is not often, and perhaps never, a constant, for the tempo of culture change varies with time. The forces for change will predominate for a considerable period and rapid alterations in the nature and structure of a society will occur. Then there may follow a period of relative quiescence, in which the unusual stresses and strains occasioned by rapid change are relieved, and the elements of the culture regroup and accommodate themselves in a more harmonious fashion. Hence, at a particular moment in time, the relative strengths of a culture's stability and proneness to change reflect the extent of the balance between the opposing forces.

Anthropologists, and increasingly technical specialists, understand a good deal about the dynamics of change in the communities in which they work. But in studying and carrying out programs of directed change, we usually forget that we are dealing with *two* societies, and not just one. For, as we shall see in Chapter 9, the administrative organizations, the bureaucracies that are charged with developmental programs, are also very real societies with equally real cultures. Their dynamic processes are astonishingly similar to those of "natural" communities. Every bureaucracy is host to both progressive and conservative forces, and, just as with a peasant community, there is a perpetual struggle to determine the pattern of slow advance. The most successful guided technological development occurs when program planners and technical specialists recognize the two sociocultural systems within which they work—their bureaucracy and their target groups—and have some understanding of the processes of change that characterize both of them. In theory, at least, the strategy of promoting change in either is simple: The strength of the conservative forces must be weakened, or their results neutralized, while simultaneously the change forces must be strengthened. Chapters 5 through 8 deal with the more common forces promoting the status quo and those promoting change in target communities; Chapter 9 deals with the dynamics of bureaucracy and the problems of technical specialists.

In studying cultural dynamics I have found it helpful to think of the change-inhibiting factors in all societies as *barriers.* There is no equally good word for the change-promoting factors; sometimes I refer to them as "stimulants to change," sometimes as "motivations to change" (when the phenomena are largely psychological), and at other times simply as "factors promoting change." Some barriers can be conceptualized primarily in cultural terms: the basic values of the group, its conception of right and wrong, the nature of the articulation of the elements of the culture, the "fundamental fit" or integration of its parts, and the overriding economic limitations that can be identified. These barriers are culture based; for simplicity's sake they are called "cultural." Other barriers are found in the nature of the social structure of the group: the prevailing type of family and the relationships among its members, caste and class factors, the locus of authority in familial and political units, the nature of factions, and the like. These barriers are society based, called simply "social." Still other barriers are most easily comprehended if phrased in psychological terms: individual and group motivations, communication problems, the nature of perception, and the characteristics of the learning process. These are psychologically based, or "psychological" barriers.

The factors that facilitate change, the *stimulants*, also can be examined in cultural, social, and psychological terms. They are the antithesis of barriers, the opposite side of the coin; and only in an analytical sense can they be separated from them. Barriers and stimulants are part and parcel of the same process and must be manipulated simultaneously. To illustrate, the values of a culture may be such that an energetic farmer easily perceives advantages to himself in adopting new cultivation practices. Whether he actually adopts these practices may be determined by familial structure and traditional reciprocal relationships within the kin group. If he is a member of an extended family in which the fortunate share generously with the less fortunate or improvident, he may feel that increased income will mean no personal gain, but rather will mean only increases in familial obligations. In this circumstance he is apt to reject the proffered aid. On the other hand, in a changing society in which such mutually accepted responsibilities are losing weight, the farmer may decide that enough of the increased income will remain with him to justify changed practices.

Cultural, social, and psychological barriers and stimulants to change exist in an economic setting. In a more comprehensive analysis, economic factors should receive extensive treatment, for they seem to set the absolute limits to change. As will be seen, people often are unwilling to change their ways because of cultural and social and psychological factors. But equally as often, they are quite aware of the value of change and anxious to alter their traditional ways, but the economic sacrifice is too great. If an economic potential does not exist or cannot be built into a program of directed change, the most careful attention to culture and society will be meaningless. Several examples will illustrate the relation between economics and other aspects of culture change.

In Tzintzuntzan, Mexico, for the past 15 years a government-operated public health clinic has attracted patients, particularly for prenatal and postnatal services and for the dressing of injuries and other similar ailments. Service is either free or offered at a nominal cost, depending on type. The cultural, social, and psychological barriers involved are, little by little, being overcome. In addition, a variety of private medical services are available ten miles away in the town of Pátzcuaro. Frequent bus service permits increasing use of these services. But many people who are convinced, or nearly convinced, of the value of medical care hesitate to take the plunge because of the cost, or fear of the cost, which is unknown and unpredictable. An infant falls ill and does not respond to folk medical care. This presents a dilemma. The parents know that the town physicians sometimes have cured children when local curers failed. But the cost is certain to be high by their standards: perhaps 50 pesos, perhaps 200 pesos. There are other small children who are not yet economically contributing members of the family. They require food, clothing, and many other things. Their well-being and that of other members of the family may be jeopardized if the parents undertake an unknown and potentially costly treatment for the critically ill infant, who perhaps, because of age, is hardly yet considered to be a real member of the family. The social cost of trying to save the tiny infant is weighed against the total family welfare, and it may be deemed too great to be justified. Furthermore, in addition to money costs, the parents will lose much productive time by the bus rides, long waits in the doctor's office, and perhaps the necessity to remain for several days or longer in a hospital.

A similar example has been described in Ceylon where from 1916 to 1922 the Rockefeller Foundation carried out a demonstration hookworm-eradication program. Although a considerable measure of success was ultimately achieved, unforeseen cultural, social, and economic factors greatly handicapped the work. Hookworm treatment is not particularly painful, but the purges weaken the patient for several days, and it was found that in the more isolated villages a major resistance to treatment arose out of poverty.

> It took some time for the field doctors doing the village work to understand the basic poverty. They struggled to overcome the evasions and refusals, and finally learned that the real trouble was that the villagers felt they could not afford the time away from work for treatment and recovery from treatment. Usually they were too weak to work for a day or two after taking the medicine (Philips 1955:288–289).

Loss of this income counterbalanced the little-understood advantages of freedom from hookworm.

Oberg and Rios have analyzed a community-development project in a small Brazilian village and have shown how social and, particularly, economic factors have affected adversely the generally sound planning. To illustrate, a latrine program was instituted as the keystone of environmental sanitation. Slabs for the pits were cast and given free to villagers, who then had to dig the hole and erect the shelter. Few people, however, took advantage of this aid, and most of the latrines eventually were installed by project workers. Analysis revealed the reasons for this puzzling lack of interest. A census showed that the village was highly unstable in terms of social organization. Half the inhabitants had lived there for less than five years and did not really consider themselves permanent members of the community. After saving a little money they hoped to migrate to other parts of the country where better opportunities might exist. Consequently they felt little attachment to the village, no stake in its future, and had no interest in making capital improvements in something they didn't expect to enjoy very long. Many people lived rent-free in their shacks, simply caring for them for absent owners. Since they might be evicted at a moment's notice, they saw little advantage in working on property that by virtue of improvement would be more attractive to the owners or to someone who might pay rent. The owners themselves, since they were away and since they received no income, had no incentive to make improvements in their homes. So the failure of this program lay not so much in the inability of the people to understand and appreciate the hygienic advantages of latrines as in social and economic factors that the planners had not understood (Oberg and Rios 1955).

The economic reality facing individual families in small communities is echoed at national levels. In rural areas particularly there is often a tendency to think of change and development as a grass-roots, bootstrap operation in which, with skillful technical help in agriculture, health, and community organization, great strides can be made by utilizing local resources, especially labor. In reviewing the history of international technical assistance we encounter also a firm belief that the highly trained technician, working both at the operational level and as a consultant at the national level, is the key to success. Yet it should be clear that without adequate infusions of capital to accompany advice and knowhow, achievements will be restricted. Maddox has made this clear in discussing the reasons for success of the well-financed Mexican antimalarial campaign of the mid-1950s:

> This project illustrates a point that may be applicable to other types of activities, namely, that substantial capital grants combined with technical assistance in organizing and managing public services may be many times more effective in solving some of the immediate problems of underdeveloped countries than the introduction of technical knowledge alone (Maddox 1956:17).

Mexico, he pointed out, had a great many highly competent health specialists, and the country needed capital more than advice.

In the discussion that follows I neglect economic factors in large measure, not because I think they are unimportant—just the opposite is the case—but because the problems of supplying capital, of loans and grants and investments, are different from the problems of rural change, and the relation between technical specialists and target groups, which is the focus of this book.

# 5

## cultural
## barriers
## to change

Within the major categories of barriers to change—cultural, social, and psychological—the specific examples fall in subgroupings. There is no magic or inner logic in the system of classification used in this and the following chapters. It is simply one author's attempt to group examples in such a way that the underlying themes will be apparent to the reader. Cultural barriers, for example, seem to fall easily into the groupings of "values and attitudes," "culture structure," and "motor patterns."

### VALUES AND ATTITUDES

#### Tradition

Some cultures value novelty and change positively, for their own sake. The fact that something is new and different is sufficient reason to examine it and perhaps to try it. Americans, we know, are attracted by the new. Advertisements play upon the theme of "new," "better," "improved," and the customer buys. In general, the positive attraction of the new and novel seems to be associated with industrial societies. Whether peoples with the most interest in novelty became the first industrialists because of this interest, or whether an industrial system produces these values, we cannot be sure. I suspect the latter—that aspirations are developed through the opportunity to satisfy them. Francis Bacon's views on change in England during its preindustrial era suggest this. While he recognized the importance of change—"every medicine is an innovation, and he that will not apply new remedies must expect new evils" —the weight of his advice cautions against giving up the old. Remarkably, this caution is stated in straight anthropological terms, that cultures are integrated systems, not to be lightly torn apart:

> It is true that what is settled by custom though it be not
> good, yet at least it is a fit; and those things which have long
> gone together, are, as it were, confederate with themselves;
> whereas new things piece not so well; but, though they
> help by their utility, yet they trouble by their inconformity;
> besides, they are like strangers, more admired, and less
> favoured (Whately 1857:225).

It is good, he felt, "not to try experiments in States, except the necessity be urgent, or the utility evident," and that "novelty, though it be not rejected, yet be held for a suspect" (Whately 1857:226). Time . . .

> is the greatest innovator. . . . It were good, therefore, that
> men in their innovations, would follow the example of
> time itself, which indeed innovateth greatly, but quietly, and
> by degrees scarce to be perceived; for otherwise, whatsoever is
> new is unlooked for—and ever it mends some, and
> [im]pairs others (Whately 1857:225).

Whatever the origin of a love for the new, the relationship between a productive economy and a tradition for change is so close that it cannot be thought of as due to chance. In contrast, in most nonindustrial parts of the world novelty and change have less positive appeal. Rather, the individual is conditioned to view new things with skepticism and, if he is uncertain, not be tempted. The great Spanish lexicographer Covarrubias, for example, defined *Novedad* (novelty) in 1611 as "something new and unaccustomed." Then, unconsciously injecting the value judgment of his society, he added, "characteristically it is dangerous because it sullies traditional usage (Covarrubias 1611:831). And nearly one hundred years earlier, in

1531, Guevara admonished the President of Granada: "Do not attempt to introduce new things, for novelties bring in their train anxieties for those who sponsor them and beget troubles among the people" (Menéndez Pidal 1950:133). Even today, in Spain and Spanish America, one of the most widely quoted of all proverbs is "*Vale más lo viejo conocido que lo nuevo por conocer,*" that is, what is old and known is worth more than something new yet to be understood.

In peasant society conservatism appears generally to be culturally sanctioned. In *The House by the Medlar Tree* the Italian novelist Giovanni Verga describes a Sicilian fishing village in the latter half of the nineteenth century. 'Ntoni, the patriarch of his family, was old, wise, and respected; he embodied the folk wisdom of his group.

> *Old Master 'Ntoni remembered many sayings and proverbs*
> *that he had learnt from his elders, because, as he said,*
> *what the old folks said was always true. One of his sayings was*
> . . . *"Be satisfied to do what your father did, or you'll come*
> *to no good."* And he had many other sensible sayings as well
> *(Verga 1955:3; my emphasis).*

This traditional wisdom is reiterated in the sayings of peasant peoples in many parts of the world: in Thailand, "If we follow the old people, we will not be bitten by the dog" (communicated by Dr. Yutthana Suksamiti); in Egypt, "He who leaves his past gets lost" (communicated by Miss Hala Nasharty); and in Sicily: "Listen to old people, for they do not deceive you" (Chapman 1971:47). Pierre Bourdieu describes similar attitudes among Algerian peasants:

> *The future is not robbed of its menace unless it can be*
> *attached and reduced to the past, until it can be lived as a*
> *simple continuation and accurate copy of the past. "Follow the*
> *road of your father and your grandfather"; "If you resemble*
> *your father you cannot be accused of fault."* Such are the
> teachings of wisdom *(Bourdieu 1963:70; my emphasis).*

It is clear that in societies where the positive strictures against being tempted by novelty are strong—where aphorisms and maxims are quoted to validate tradition and where fear of criticism haunts the would-be innovator—a fertile field for a broad program of social change does not exist until after a good deal of preliminary cultivation has been done.

## Fatalism

The attitude of fatalism is closely allied to the forces of tradition and constitutes a barrier of equal strength. In industrial societies people have proved to their satisfaction that a high degree of mastery over nature and social conditions is possible. An undesirable situation is not a hopeless block, but rather a challenge to man's ingenuity. In industrial societies people have come to believe that almost anything can be achieved; at least, any reasonable plan is worth a serious try.

But in nonindustrial societies a very low degree of mastery over nature and social conditions has been achieved. Drought or flood is looked upon as a visitation from gods or evil spirits, whom man can propitiate but not control. Feudal forms of land tenure and nonproductive technologies may condemn a farmer to a bare subsistence living. Medical and social services are lacking, and people die young. Under such circumstances it is not surprising that people have few illusions about the possibility of improving their lot. A fatalistic outlook, the assumption that whatever happens is the will of God or Allah, is the best adjustment the individual can make to an apparently hopeless situation.

The Colombian anthropologist Virginia Gutiérrez de Pineda studied cultural factors involved in the high rate of infantile mortality in rural areas of her country, and she points out in poignant fashion the lethargy that social and economic conditions have forced upon the countryman. When an infant dies, the parents say, "It was his destiny, not to grow up." In Santander province it is often said of an unusually beautiful child, "This child is not for this world," and thus the parents prepare themselves for the 50 percent probability that the child, in fact, "is not for this world." On the other hand, when an ill child recovers, the parents say, "See, he recovered without medical attention; God did not intend him to die." In the face of such attitudes, Pineda found that the trained physician had a difficult time in gaining the confidence of the people. When she would urge parents to take an ill child to see a doctor, often they shrugged their shoulders and replied, "The rich also die, in spite of having so much money for medical care" (Pineda 1955:18–19).

A fatalistic attitude is widespread in rural Latin America. In the Brazilian village of Cruz das Almas, near São Paulo, described by Pierson, it is commonly believed that illness comes from God, having been sent, often, as punishment for sins, and that it is "He who either cures or 'takes you away to the other world.' . . . In the event of illness or death, one often hears the phrase '*Deus quis*' (God willed it). If the illness is prolonged, this fact is considered a part of the person's *sina* (destiny)," and characteristically one asks the rhetorical question, "What is there to do about it?" (Pierson 1955:281–283).

Religious beliefs and sacred texts often contribute to fatalistic attitudes. In the parched farmlands of northeastern Brazil a health worker found it was especially difficult to persuade mothers to seek help for their sick children during the month of May. This is because in Catholic teachings May is the "Month of the Virgin Mary," and in this part of Brazil it is believed that when a child dies in May it is particularly fortunate, since the Virgin is "calling" her children to come to be with her. "To seek medical aid during this time would be to contravene the will of the Virgin" (communicated by Marina Beatriz Cruz Santos). In Egypt death is perceived as Allah's will, and no one can extend life because the Koran says "Wherever you are, death will seek you, even if you are in strongly built castles." An Egyptian physician sees this attitude as one of the reasons for high infant mortality in his country (communicated by Dr. Fawzy Gadalla). Egyptian proverbs also teach fatalism, as the following illustrates: "Run as fast as wild animals and still you will not get more than what is assigned to you by God" (communicated by Zeinab Shahin). A budding entrepreneurial spirit is hardly to be encouraged by this reminder.

### Cultural ethnocentrism

We Westerners, proud of our achievements in science and technology, often believe they imply that our total culture is the most advanced and is therefore superior to the cultures of simpler peoples. Our conviction of superiority and our belief that we have knowledge of truth make us anxious to "share" this superiority with other peoples whom we believe to be less fortunate. It sometimes comes as a surprise to us to discover that the members of all cultures believe that basically their way of doing things is natural and best. The Greenland Eskimo illustrate this very clearly: They believe that Europeans have come to their land to learn virtue and good manners from them. To say that a European is, or soon will be, as good as a Greenlander is the highest praise the Eskimo can bestow on him (Gittler 1949:44).

Primitive peoples are quite willing to acknowledge the superiority of a steel knife over a stone knife, and sometimes of an aluminum kettle over a pottery vessel. But these are peripheral and inconsequential areas of culture. The real essence of culture, all of us believe, lies in what we think and do, our attitudes, our social forms, and our religious beliefs. The question of superiority in such things is, of course, harder to measure or prove. Begging the question of absolute values, it is apparent that the universal belief in the superiority of one's culture is a powerful force for stability. This is as true of the American as of the Australian bushman.

The anthropologist studies ethnocentrism in terms of what he calls "cultural relativism." By this he simply means that the values of all peoples are a function of their way of life and that they cannot be understood out of context. The point of view of relativists is not that all ways are equally good—they do not endorse slavery, murder, and other conditions or acts in which the individual is deprived of rights or opportunities to realize his full potential—but they do hold that it is wrong to condemn the ways of others simply because they differ from those of the person who is passing judgment.

Ethnocentrism is so deeply engrained in all of us that even when we are sensitive to the philosophy of cultural relativism we may easily fall victim to evaluating others in terms of our own views. It is Pineda again, who so clearly points out this danger. Her field work in Colombia has included study of the lives of the cattle-raising Guajiro Indians of the Guajira Peninsula.

> I remember when once I spoke with an Indian woman of
> high social class about marriage, and the Indian custom of
> giving money and cattle to buy the wife. I had not yet
> come fully to understand the Indian culture, and while the
> woman spoke of her price I felt terribly sad that a Colombian
> woman could be sold like a cow. Suddenly she asked,
> "And you? How much did you cost your husband?" I smugly
> replied, "Nothing. We aren't sold." Then the picture changed
> completely. "Oh, what a horrible thing," she said. "Your
> husband didn't even give a single cow for you? You must not
> be worth anything." And she lost all respect for me, and
> would have nothing further to do with me, because no one had
> given anything for me (communicated by Virginia Gutiérrez
> de Pineda).

In contemporary Africa similar views have been expressed. Younger men are beginning to look upon the traditional "bride price," which they must pay to the girl's father, as blackmail, since it may represent the equivalent of up to five years of income. Young women seem less anxious to change the status quo. At a YWCA conference in Uganda one young delegate asked, "How will our husbands value us unless they have given value for us?" And another wondered, "How can our husbands keep us faithful unless there is a dowry they can demand back?" (*Time*, July 30, 1965).

## Pride and dignity

Anthropologists have noted that an innate dignity in personal bearing and a pride in their way of life characterize the peoples among whom they work. This correlates with the ethnocentric position of most people vis-à-vis their cultures, and it is reflected in a strong—not to say rigid—belief about the behavior appropriate to recognized roles. Many technically well designed aid programs have run into trouble because culturally defined forms of pride and "face" which express these strong feelings about role have not been recognized. Desire to avoid humiliation as a result of being cast in an inappropriate role seems to be universal. But what determines propriety is determined by culture. In the United States, for example, the idea of lifelong learning is deep-rooted, and adults do not hesitate to take correspondence courses or attend night school if they feel they will profit thereby. The role of student is one that the individual may occupy at any time during his life without fear of ridicule.

But in much of the world schooling is associated with childhood. The role of student is all right for youngsters, but it is inappropriate to the adult state. Dube points out that in India, where the Community Development Programme stresses adult literacy projects, adults often quickly drop out of night classes, even though literacy and education are highly valued. Since only children are supposed to attend schools, the adult exposes himself to general public amusement when he starts to school with pencil and slate (Dube 1958:122).

Fear of loss of face can threaten agricultural programs also. One phase of the Indian agricultural program includes selling improved seed to farmers at moderate cost. Curiously, the best and most progressive farmers often are the most resistant to this aid. In one village the wealthy and able farmers neither purchased nor used this seed. "It has long been thought a disgrace and a sign of failure or poor management to be forced to borrow or buy seed. The village farmer takes special pride in being able to raise enough food to maintain his family and in having enough left over to use as seed" (Opler and Singh 1952:7). The able farmer does not wish to be cast in the role of the incompetent agriculturalist.

Isabel Kelly tells that in parts of Mexico it has been difficult to persuade people to use a yellow maize that is superior nutritionally to the local white variety. The explanation is found in the locally accepted evidence of good cooking. The tortilla, the staple of life, is made by soaking hard maize grains in lye water, grinding a dough, patting out thin circular cakes, and baking them on a clay griddle. Tortillas made with white maize are white when cooked, unless too much lye has been added, which turns them yellow. In the villages described, yellow tortillas resulting from the use of yellow maize were identified with careless cooking, and housewives did not wish to be stigmatized as careless or incompetent cooks (Kelly 1958:205–208). Kelly found a comparable problem in Bolivia during her work as a public health consultant. The native *quinua* is an extremely nourishing grain, with more protein, fat, and vitamin B than wheat. From a nutritional standpoint it would be desirable to extend its use. But because in the popular mind it is associated with the poorest classes, this proved to be difficult. Even Kelly's cook refused to serve it in the presence of guests; she did not want to compromise her culinary reputation by serving a lower-class dish (Kelly 1960:22).

Fear of loss of face has handicapped maternal and child health programs in Taiwan. There, where the extended family is still strong, the older women, in whom much familial authority is vested, feel that it reflects on their ability and judgment if young pregnant women in their families attend prenatal clinics or seek the aid of trained midwives. Young women who have been convinced by a visiting public health nurse of the value of new practices have been prevented from adopting them because of the authority of their elders (communicated by Merle S. Farland, R.N.).

In the preceding chapter it was said that part of the strategy of directed culture-change consists in identifying barriers and then seeking ways to weaken them or neutralize their effect. The two following examples show how different problems were successfully attacked.

The Organization of American States operates the Inter-American Housing Center near Bogotá, Colombia. Young architects and other potential urban planners come from many countries to learn the new science of city and regional planning. Firsthand knowledge of materials is believed by the staff to be an important part of the training, and students are asked to mix mortar, lay bricks, and otherwise engage in manual labor. Most Latin Americans, however, have doubts about the dignity of working with one's hands. A professional man uses his mind, not his brawn; hired peons do the menial work. For this reason the work part of the curriculum was not successful in the early stages. Then someone hit upon the idea of supplying the students with white laboratory coats with their names embroidered over the pocket. In the new and acceptable professional status of laboratory technicians they now happily went about the chores they formerly had resisted (communicated by L. Currie, former director of the Center).

The other instance comes from Chile where public health centers, modeled after United States patterns, were introduced beginning in the 1940s. The prenatal mothers' "class" taught by a public health nurse was a part of the introduced pattern. But the new program was only partially successful; expectant mothers balked at being taught in classes like children. Consequently it was decided to represent the classes as short-lived "clubs," which met for the prescribed number of weeks, usually in the homes of the mothers. The health center provided tea and cakes, and the meeting thereby became a social affair, in which the discussion of prenatal care was only an incidental event. Since club life is associated with the upper and middle classes, the women from low-income brackets who were health center patients were delighted to be asked to participate in such activities, and the program, as a health measure, has been highly successful (author's field notes).

### Norms of modesty

Like feelings about dignity and pride, the ideas as to what constitutes modesty are instilled in the members of all societies by their cultures. These ideas are culturally defined and differ greatly from one community to another. Proper behavior in one group may be shockingly improper in a second. But no culture is without the concept of modesty. Most often such ideas seem associated with dress and decorum in behavior. Modest dress more often than not is associated with covering of the sex organs, but this is by no means universal, and among many peoples nudity or seminudity is taken for granted. Travelers returning from the Amazon Basin tell of the embarrassment of Indian women who were observed at close range in a state of nature. This embarrassment usually disappeared when the women had retired a moment to reappear wearing a string of beads or some other simple ornament. In Mexican villages a housewife is greatly upset if a caller surprises her without her apron, even though she is wearing a skirt and numerous petticoats. And older men are most uncomfortable if seen without a hat.

In programs of directed culture-change, ideas of modesty often constitute serious barriers to some kinds of programs. For example, widely prevalent ideas about female modesty and the proper relation of a physician to a pregnant woman have seriously handicapped medical and public health workers in their efforts to reduce infant and maternal mortality. In Moslem countries, in Latin America, and in many other areas, it is quite unthinkable that a man other than a woman's husband should have the degree of intimacy with her required by a gynecological examination. The impersonality of modern medicine, largely taken for granted in Western countries, is not a part of the understanding of most of the rest of the world, and many women prefer to avoid scientific prenatal care rather than to submit to examination by a male physician. Often, too, it is the husband who objects to this routine (to Westerners) treatment. This difficulty usually can be overcome by using female physicians, but here again the problem is great. Until very recent years women have not been permitted to study medicine in most of those countries in which this resistance is found, and it will be many years until there are sufficient female physicians to meet the demands of adequate prenatal services.

Although female physicians probably are the answer in most parts of the world where modern medicine in all its aspects has not been accepted, this is not a blanket rule. Schneider points out how women on the island of Yap in Micronesia are resistant to genital examination by a male physician, but they are even more resistant to examination by a female. Yap women regard all other women, regardless of age, as potential rivals for men's attention; at the same time, they believe their own genitals are their strongest power over men. Exposing their source of power to potential rivals, they believe, weakens their competitive position and threatens them with loss of masculine attention (Schneider 1955:231).

Among the Navaho there is not too much resistance by parturients to male physicians; male assistants have long helped with childbirth under native conditions, and it is not thought particularly shameful for men to be present at such an event. But resistance to hospitalization for delivery is strong, largely because of the customary short hospital gown. Navaho modesty requires good coverage from waist to ankle for women, and the American hospital gown is offensive in the extreme (Bailey 1948:1420).

## Relative values

There is a tendency for specialists who work in technical aid programs to assume that the people with whom they are working, and whom they hope to help, are essentially rational, however uneducated they may be and however simple their way of life. It logically follows that if the technician has difficulty in putting his program across, either the people are unusually stupid and can't see the obvious (to the specialist) advantages of change, or the technician has not been as skillful as he should be in the presentation of his case.

In fact, people often understand the message perfectly, but they have weighed the relative values to them of the alternative forms of behavior and have decided against the new program, or some part of it, however compelling the evidence may seem to someone with a different point of view. The technician, frustrated by some example of irrationality on the part of the people he wishes to help, may attempt to relieve his tension by furious chain smoking, even though he knows the medical evidence for a correlation between the use of cigarettes and lung cancer is pretty convincing. He has made a value judgment and prefers the satisfaction of smoking to the probability of lessening his susceptibility to cancer.

The satisfactions of life are many and varied, and economic rationality, though very important, is far from being the only determinant of value judgments. The agricultural extension agent, struggling to solve problems of food shortage, often wonders why people are reluctant to grow more nutritious and higher-yielding strains. One of the most striking examples of this situation has been summarized by Apodaca. In a community of Spanish-American farmers in the Rio Grande Valley of New Mexico, a Department of Agriculture county extension agent succeeded in 1946 in introducing hybrid corn that produced three times the yield of the traditional seed. After participating in initial test demonstrations, a majority of the growers adopted the innovation. Yet four years later nearly all farmers had reverted to the old corn. Investigation revealed that the farmers' wives had complained about the texture of the dough used to make tortillas, about the color of the finished product ,and about the taste. In the system of values of this community, corn quality turned out to be more important than corn quantity; people were willing to sacrifice economic gain for something they esteemed, in this case traditional food characteristics (Apodaca 1952).

This is not an isolated example. Eating habits are among the most emotionally based of all activities, and unfamiliar taste often turns out to be a reason why new foods are rejected, unless they can be cloaked with even more powerful counterarguments, such as prestige value. In India a village worker asked a farmer about his reaction to a new wheat that had been tried experimentally for several years. The villager replied that it was better in appearance, brought a higher price, and was resistant to rain and frost. Then he added, "But the local variety is better in taste," and concluded, "From the point of view of health there is nothing like it (Dube 1958:196). In 1957 I encountered the same story in the Helmand Valley in Afghanistan. Improved maize seed had been introduced, which, when planted in rows (rather than broadcast) and cultivated with improved techniques, produced up to three times the yield of local varieties. Although the new seed was slowly making progress, resistance to it was surprisingly strong. One of the principal reasons given by farmers for their lack of interest was that bread baked from the new maize compared unfavorably in taste with bread made with the traditional maize.

## CULTURE STRUCTURE

### Logical incompatibility of culture traits

Not all culture elements or institutions can be easily combined. Between some there is a logical compatibility, between others, a logical incompatibility. When logical incompatibility exists, change comes about with difficulty. The contrast between a monotheistic and a polytheistic religion illustrates this point. Peoples who practice polytheism often are not unreceptive to Christian missionaries. It seems obvious to them that this new god is a powerful diety, or his representatives would not be so far from their homes; he appears powerful, also, because of the power he has delegated to his lieutenants. When these lieutenants ask them to accept him, often they are able to do so with great ease and a clear conscience, because no serious violence is done (at least in the early stages of Christianization) to their basic religious beliefs. They already pay homage to a number of dieties with special attributes; they have added dieties from time to time in the past; and it is the most natural thing in the world to add a new one who obviously possesses special merit.

On the other hand, proselytizing among monotheistic peoples offers a much more difficult problem. Monotheistic peoples can accept a new diety only by rejecting the previous incumbent, or by completely identifying the new god with the old one through the process of syncretism. This is asking a good deal more of people than is necessary with polytheistic groups.

The contrast between autocracy and democracy is similar. By definition, autocracy cannot permit important divergencies of opinion, but democracy does not exist unless these divergencies are present. The two forms are logically incompatible and cannot be reconciled.

Even on a less sweeping level logical incompatibility of culture elements poses barriers to change. For example, the Navaho Indians have resisted both Christianity and pagan nativistic movements because their religious beliefs are incompatible with those offered by the alternate forms. They were immune to the Ghost Dance movement of 1890, which swept many western reservations and which taught the resurrection of the Indian dead, the removal of the whites, and the reestablishment of the old order. W. W. Hill suggests that "for the Navaho with his almost psychotic fear of death, the dead and all connected with them, no greater cataclysm than the return of the departed or ghosts could be envisaged. In short, the Navaho were frightened out of their wits for fear the tenets of the movement were true" (W. W. Hill 1944:525). Reichard believes the Navaho abhorrence of death makes it very difficult for them to comprehend Christianity, which is based on the concept of death and resurrection (Reichard 1949).

In a similar vein, Hawley explains Hopi resistance to Catholicism as due to the logical incompatibility between the structures of the two systems. Catholicism, she points out, is a highly centralized religion, patriarchal and patrilineal in form. Among the Eastern Pueblos of the American Southwest, religious organization and basic social structure (even in matrilineal instances) are quite similar to the Catholic pattern; therefore these people were hospitable to the new religion. But the Hopi rejected the new religion, and killed priests and burned their missions. Hawley believes that matrilineality coupled with decentralized governmental forms explain why the Hopi have maintained their religious independence (Hawley 1946).

While the logical incompatibility of traits is usually thought of as characterizing whole cultures, it may apply also to individuals—technical specialists as well as natives—in extreme cases making the specialists' jobs impossible. In 1965 in Nepal I encountered the case of a young Brahmin agricultural extension agent assigned to a Rai tribal village where, after hard work, he gained the confidence of the people. As evidence of their friendship they invited him to drink the local distilled liquor, called *rakshi*. But as a nondrinking Brahmin he felt he could not do so. The villagers, interpreting his refusal as a rejection of their friendship, ignored him, and he had to be recalled and assigned to another village.

## Unforeseen consequences of planned innovation

No change can occur in isolation. Any induced change will produce secondary and tertiary changes over a wide area. It is a little like throwing a pebble into a pond. The effect of the blow on the water will extend in ever-widening circles until the impact loses its force. In the same way, the impact of an introduced trait will extend in widening circles until its effect is felt in areas of culture far removed from the point of contact. Conversely, the extent to which changes can be introduced depends on other changes that are taking place, or can be made to take place, which will influence the reception accorded the innovation.

Of all of the practical guides anthropology can offer to technical specialists, none exceeds in importance the necessity to consider carefully all possible consequences of a new program. Most culture change involves a social cost. However logical and desirable an innovation may appear to the scientifically trained technical specialist, some of its secondary and tertiary consequences may be highly undesirable from the standpoint of the people affected, and apparent advantages must be weighed against possible disadvantages.

A striking example of failure to consider all of the consequences of technological "progress" comes from Israel, where the beautiful Sea of Galilee is in grave danger of becoming a gigantic pool of sludge. Much of the pollution flowing into the sea consists of untreated sewage, agricultural chemicals, and industrial wastes. But the single largest source of pollution is the unanticipated price of success in draining the Hula Swamp, 15 miles to the north. Drainage work, completed in 1957, eradicated the scourge of malaria from Galilee and provided thousands of acres of fertile land. But nitrates from peat in the former swamp now flow unchecked into the sea and today constitute 40 percent of all polution. The situation is so serious that some experts even suggest the reflooding of the Hula Swamp! (Meyers 1971).

Taylor has pointed out the high social costs of a technically successful agricultural project in Haiti. The Artibonite Valley dam provides irrigation waters desperately needed to increase food supplies. At first glance this might seem like a bonanza to the local farmers who owned their small plots of land, which gave them at least a minimum degree of security as well as equality with their fellows. But as land is improved it becomes increasingly attractive as a capital investment for city dwellers. Illiterate peasants, dazzled by what are to them fantastic offers for their land, sell and are quickly reduced to the status of landless proletariat. Although some of them continue to work for wages, high land capitalization places a premium on larger units that can be farmed with machinery, thus reducing the demand for hand labor. So, for the Artibonite Valley peasants a notable engineering achievement may well be an unmitigated disaster (Taylor 1954).

One of the primary tasks of the anthropologist working with developmental teams is to follow through on the possible consequences of a proposed change, to analyze the factors involved, and to attempt to predict what will happen. If he knows the culture well and if he is familiar with the general processes of culture change, he should be able to anticipate unforeseen reactions that can seriously affect the outcome of planned change. Often, if these critical points can be dealt with in planning, they can be neutralized; if they are ignored, failure is apt to result. The following examples show how proposed innovations must be appraised in broad perspective rather than in terms of a single goal.

In village India, cooking is traditionally done over an open dung fire in the kitchen. There is no chimney and there are few windows, so the room fills with choking smoke, which gradually filters through the thatch roof. Cooking is unpleasant under such conditions, and respiratory and eye ailments are common. The Community Development Programme has recognized this situation as a serious threat to health and has developed an inexpensive pottery stove, a "smokeless chula," which maximizes the efficiency of fuel and draws smoke off through a chimney. It is sold at very low cost to villagers. Yet the smokeless chula has had limited success. In much of India woodboring white ants infest roofs; if they are not suppressed they ruin a roof in a very short time. The continual presence of smoke in the roof accomplishes this end. If smoke is eliminated roofs must be replaced far more often, and the expense is greater than farmers are able to support. So the problem of introducing the smokeless chula —at least in many areas—lies not in the villager's addiction to smoke-irritated eyes, nor in his love of tradition, nor in his inability to understand the cooking advantages of the new stove, nor in the direct cost of the stove itself. He has considered the trade-off alternatives and decided that the disadvantages of the new stove outweigh the advantages. In this case the critical area of resistance has nothing at all to do with cooking.

Renteln describes a similar example in Iran. On the south shores of the Caspian Sea, smoke-filled houses cause health problems similar to those found in India. Yet resistance to better cooking methods was found to be intense. Here, too, smoke has its positive side. The area is heavily infested with anopheles mosquitoes and malaria is widespread. A high level of smoke in the house is the only way the villager can obtain relief from bites (communicated by Dr. Henry Renteln).

These examples illustrate an important point in the strategy of directed culture-change. If the reasons for resistance are analyzed, it may be found that a series of innovations that tie together and mutually reinforce each other will make success possible, where single innovations will fail. The Indian villager probably did not know, before he tried the smokeless chula, that smoke preserved his roof. He discovered it when smoke was removed. If the roof can be preserved by other methods, the threshold of his resistance to improved cooking methods will be greatliy lowered. Malaria control with regular pesticide spraying is an important project of the Community Development Programme. If spraying is co-ordinated with the demonstration of the smokeless chula, white ants as well as mosquitoes will be eliminated, and one of the costs—economic in this case—that constitutes a barrier is eliminated. Similarly in Iran the same double program should go far in solving the problem of the villagers' resistance.

Other examples illustrate barriers that may be lessened if they are approached, not in isolation, but as part of a cultural complex, which must be examined as a unit. Good animal husbandry prac-tices require the castration of many male domestic animals: pigs and sheep so treated grow fatter, and oxen are larger and stronger. The scientifically minded animal husbandryman sees these advan-tages as functions of farming practice in his own country, and the advantages seem to far outweigh the disadvantage of an occasional infection or lost animal. Yet this practice often meets strong opposi-tion in newly developing areas. In 1957 I found great resistance to castration of young bulls in the Helmand Valley of Afghanistan. The principal reason was entirely logical from the standpoint of the farmer: Fields are tilled with a crude wooden plow fastened to an equally crude yoke, which rides on the neck of the pair of oxen. The yoke is held in place by the large hump that develops on the shoulder of the mature animal. If bulls are castrated while young—the proper time if maximum benefits are to drive from the operation—this hump does not develop, and the animal is useless for traction. Y. Subrahmanyam tells me that among the

Koya hill tribes of southeast India he encountered similar resistance, and the same reason (communicated by Dr. Y. Subrahmanyam).

It is obvious that if castration of bulls is to be a serious part of a livestock program in these areas, a new type of yoke must be introduced. This should not be too difficult. The Mediterranean and Latin American yoke is quite similar and no more difficult or costly to manufacture. It is lashed to the horns of the oxen, thereby eliminating the need for a hump.

In recent years, as a part of the United States technical aid program, attempts have been made to introduce the native American bean *(Phaseolus vulgaris)* into the Helmand Valley. This crop often proves to be an inexpensive and effective way to introduce needed protein into a diet marked by meat deficiency, and it appeared to be a logical step in Afghanistan. Although the local people liked the new food and it grew well, resistance ran high. It was then discovered that a shortage of fuel places a premium on quick-cooking dishes; the bean was uneconomic in this sense. Subsequently, the black-eyed pea, which has high nutritive value but cooks in much less time, has been rather successful.

In villages in the high Andean area of South America a fire smoulders in the middle of the floor of one-room houses night and day. In programs of home improvement attempts have been made for sanitary reasons to get housewives to use a raised fireplace. These attempts have been largely unsuccessful. The Ecuadorian anthropologist Aníbal Buitrón points out that in these home betterment programs technicians viewed the fire as simply a cooking device, whereas it also serves the equally important function of affording warmth to householders, who sleep around it at night and huddle around it when they come indoors on cold days. A sanitary, raised hearth appreciably reduces the value of the fire for heating. Buitrón suggests that the dual function of the fire might successfully be met by introducing the corner fireplace of the pueblo Indians of the southwestern United States. This would reflect heat into the room and at the same time give some degree of protection to food and dishes (Buitrón 1959).

Immediate answers to all linked problems are not always apparent. Another example from the Helmand Valley points up the sensitivity of culture to a single change. The local oxen are thin and scrawny and have relatively little strength. The traditional plow merely scratches the surface of the earth to a depth of a few inches and does not turn a furrow. Moldboard plows, which cultivate much more effectively, have been tried, but the animals are too weak to pull them. In some of the newly irrigated areas of the valley better pasture has made possible fattening of oxen, which seem to meet this drawback. But the fattening introduces still another unforeseen difficulty. The well-fed animals are much less docile, do not obey commands well, and are generally unmanageable with customary control techniques. Consequently, many local farmers are not anxious to feed their animals too well. A weak, tractable draught animal is preferable to a strong, uncontrollable one. More perfect domestication, or better methods of harnessing and driving oxen must be developed before this problem can be met successfully.

The two following examples illustrate strikingly similar unforeseen consequences resulting from failure to think through the possible results of agricultural innovations. In 1951 a yellow Cuban maize was introduced into the eastern lowlands of Bolivia in the Santa Cruz and Yungas regions. It had many apparent advantages. It grew well in the tropics, matured more rapidly, had more fat content than local varieties, was less subject to insect attack, and produced a higher yield per unit of land. The new maize seemed to be an excellent device to improve the diet of both people and animals, and it was for this reason that it was introduced. It has proved very popular, but not for the reasons anticipated. Its very hardness, desirable from the standpoint of storage, makes is difficult to grind, and people are unwilling to take the time and trouble to haul it to commercial mills in towns. But it makes an excellent commercial alcohol, and prices are high. Thus, a seemingly desirable innovation has promoted alcoholism instead of improved diet (Kelly 1960:10–11).

Attempts to substitute cassava for millet in Northern Rhodesia (now Zambia) produced similar unexpected and undesired results. Millet is not a very satisfactory commercial crop in this part of Africa. It is attacked by locusts and is easily affected by changes in the weather. Cassava is easier to grow and more profitable. But the economic basis is not the only one to be considered in attempting a simple substitution. In South Africa millet is essential for beer, which plays a vital role in the social and religious life of the people. Further, however unsatisfactory it may be as a cash crop, it is important in diet. Since it is rich in fat content it is an important substitute for meat and milk, unprocurable because the tsetse fly makes it difficult to raise cattle. Compared with millet, cassava is deficient in fat content, so nutritional standards were lowered by this enforced change. Perhaps worse, with the prohibition of millet beer, drunkenness increased; the natives substituted a concoction called *skokian,* composed of methylated spirits, calcium carbide, molasses, sugar, and other ingredients. The new drink was not satisfactory as a social or ceremonial substitute for the old; it was dangerous; and it aroused great hostility toward the administration (Rosentiel 1954).

In these instances the damage has been done, and backtracking for a second try is probably out of the question. In other cases the innovator may be perceptive enough to recognize the problem of the most practical, or least harmful, trade-off, even if the best answer is difficult to determine. Hazel I. Blair, an American physician working with the Alaska Eskimo, was faced with two dilemmas presented by Echinococcus disease (the cystic larval stage of a small tapeworm). This disease, prevalent on St. Lawrence Island in the Bering Sea, is transmitted through sled dogs, and effective control of the disease means extermination of the dogs. But what substitute is there for the traditional means of transportation? Or again, Blair says, "An interesting decision the doctor may have to make in areas where people subsist almost entirely on fish is whether it is better to have fish tapeworms from eating raw fish or to run the risk of a vitamin deficiency by cooking the fish." Furthermore, in fuel-short areas, how is the cooking to be done? (Communicated by Dr. Hazel I. Blair.) Obviously, a straightforward, single-minded answer to these questions may do as much harm as good.

Significant opposition to the World Health Organization's antimalarial DDT residual-spray campaign of the 1950s was based on a curious consequence. Cats, widely kept as mousers, rubbed against house walls, licked their fur, ingested the DDT, and died. In Taiwan, rat populations increased rapidly and became a serious threat to stored food (communicated by Merle Farland, R.N.). In Mexico, kittens, previously drowned or given away, came to command the astonishingly high price of $.36 U.S., so short was the supply of trained mousers (author's field notes).

Faced with some innovations, villagers have the option of refusing to go along. To development workers this resistance is frustrating, but often it is based on good reason. The villager, at least, can himself decide on the relative merits of the trade-off. In the early 1960s a Japanese rice was introduced into Nepal. Coupled with improved cultivation practices, it produced up to 200 percent more harvest than native seeds and methods. In spite of such apparent success, however, the innovation was unpopular when I observed it in 1965. First, the seeds clung so tightly to the stalks that traditional hand threshing methods were inadequate and a special mechanical threshing device had to be used, thus adding to costs. And, second, the new rice grows on a dwarf stalk that provides much less fodder than the indigenous rice. Since animal fodder is a basic by-product of rice cultivation, reduced fodder supplies presented peasants with a serious animal feed problem (author's field notes).

## MOTOR PATTERNS AND
## CUSTOMARY BODY POSITIONS

It is apparent to all of us that we learn speech patterns with ease in childhood but that as adults it is far more difficult to duplicate exactly the sounds of a foreign language. Motor patterns and customary body positions also can be thought of as a way of expressing ourselves that is easily learned in childhood but modified only with difficulty when we are adults. Culture determines the positions in which we sleep, stand, sit, and relax. Culture determines what gestures we use, how we hold and use tools, and how we manipulate our bodies in an unlimited number of situations. In Negro Africa, workers in the fields stand with straight legs, bending almost double at the waist to work with short-handled tools. The Mexican Indian also uses short-handled tools, but he squats close to the earth. American craftsmen usually work on an elevated bench, and dexterity with toes and bare feet is not essential. In many places—I have noted it both among the Popoluca Indians in Mexico and in Hindu villages—work is done on the ground, and the feet are almost as important as the hands. Americans hold a knife so that the blade projects on the side of the thumb and forefinger; Eskimos and Northwest Coast Indians prefer a knife blade to project from the opposite side of the clasped hand. American carpenters draw a saw toward them; in Latin America carpenters push a rip saw away from them.

In childhood we acquire all the patterns characteristic of our culture in the same unconscious fashion that we acquire the speech of our society. And just as the necessity or desirability of using a new language may confront the adult with difficulties, and perhaps induce him to avoid or reject the situation in which the new language would be advantageous, so the necessity or desirability of modifying motor patterns may cause the adult to reject new ways that would be of advantage to him.

To illustrate, Mexican potters use a variety of techniques and tools in forming vessels. The most advanced is, of course, the wheel. There are many skillful potters, however, who use hand modeling methods or who form pots in molds. In the state of Michoacán most of these molds consist of two halves with a vertical division; each half is a mirror image of the other. The potter places a thin sheet of clay in each half, trims the edges, and joins the sections, rubbing the weld with her hand to seal the joint. When the molds are removed the pot is perfectly formed and needs only a little polishing before firing. In this work there are few rotary or revolving motions made with the hands.

In other parts of Mexico the potter places the clay over a mold that is like an upturned bowl. This forms the bottom of a vessel. The sides of the pots are built up with coils or daubs of clay, by pinching and by turning the developing vessel round and round. Then the potter takes a wet piece of leather, and with a circular motion smooths the outside and inside. Thus, motor patterns differ substantially from those of Michoacán in that the pot rotates, even though very slowly, and the potter smooths the vessel in much the same way that a wheel-using potter smooths a vessel.

In the 1930s the Mexican government, desiring to improve the quality of pottery made in the country and to increase the income of certain villages, established trade schools in which master potters taught more modern techniques, including the use of the wheel. Research some years later revealed that in Michoacán villages, where vertical-halves molds were used and few rotating or circular hand motions were customary, the wheel was not adopted. Apparently traditional motor patterns of adult potters were so rigid that they could not or would not take the time to master a radically different technique. But in those villages using the second type of mold—the upturned bowl—in which rotary and circular motions were involved, the wheel was often accepted. Old motor patterns were speeded up, but basically new forms did not have to be learned (Foster 1948).

To have to change customary motor patterns is both difficult and tiring. As part of a community-development program in the Cook Islands, a raised cooking stove was developed so food would be protected from dirt and animals, and women would not have to stoop continually in doing their work. The raised stove generally was rejected. The complaint: It was so terribly uncomfortable to have to stand on one's feet while cooking (communicated by Mary

Hopkirk). Dube tells how an Indian village well was cleaned, bricked, and provided with a waist-high protective wall to reduce the danger of contamination. At first the villagers' enthusiasm was great, but shortly their interest waned. The high protective parapet required the use of pulleys (an innovation), thus necessitating a change in the traditional bending posture for drawing water. The women found the new set of motor patterns more exhausting than the old ones (Dube 1958:134).

Comparable lessons have been learned in environmental sanitation programs where the sanitary latrine has been introduced. American technicians have played a major, although not a dominant, role in popularizing the latrine in many developing countries. At first the outhouse may not seem to represent modern technology, but it constitutes an enormous advance, from the standpoint of health, over the ways of disposing of bodily wastes traditional in many places. Not surprisingly, the American technician thinks of the latrine in terms of the style formerly prevalent in this country: a square wooden structure with a raised seat perforated by one or more holes. Yet this type of latrine has had limited acceptance. An incident in El Salvador illustrates the problem. Several years ago a coffee planter, interested in the welfare of his employees, built a latrine for each house according to the standard American model. He was upset when his employees refused to use them. Finally an old man offered a suggestion. "Patrón, don't you realize that here we are squatters?" The planter ripped out the seats, replaced them with a perforated slab floor, and was gratified to find that public acceptance was much greater (Foster 1952:13). He had learned what has had to be learned independently, time after time, in many parts of the world: for psychological or physiological reasons, latrines with raised seats seem to cause constipation among people who customarily defecate in a squatting position.

When new tools and technical innovations can be adapted to traditional motor patterns, the probability of acceptance is greater than if no attempt at accommodation is made.

## SUPERSTITIONS

Probably all developmental workers have encountered instances of resistance to change that can best be described as due to superstition, an uncritically accepted belief not based on fact. In 1962 in Northern Rhodesia (now Zambia) I found that nutrition education efforts were hampered because many women would not eat eggs; according to widespread belief, eggs cause infertility, they make babies bald, and they cause women to be licentious. In the Philippines it is widely believed that chicken and squash eaten at the same meal produce leprosy (communicated by Dr. Narcisco R. Lapuz). In a community-development project area in India according to Dube, smallpox is classified as a "sacred" rather than a "secular" disease, the visitation of a Mother Goddess; consequently, prescribed rituals and worship receive more attention than proper medical care. (Dube 1956:23). Dube also reports pregnancy and infancy superstitions that proved resistant to change. Women were not given milk during late pregnancy and following confinement because they believe that it produces a foetus too large for easy delivery, and that it causes swelling and pus in the mother's fallopian tubes. A baby is not given water for several months after birth, because water's "cold" quality is upsetting to the infant's heat equilibrium. "The efforts of the midwives appointed by the [community-development] Project had very limited success in changing the traditional practices of the community because of the strength of conventional beliefs in this area" (Dube 1956:24).

In spite of the difficulty in coping with what to scientifically-trained people appear to be baseless superstitions, imaginative approaches sometimes overcome these barriers at least partially. In parts of northern Ghana, as in Zambia, hens' eggs are taboo to women; health education programs therefore urge the substitution of guinea-hen eggs in recommended diets. In other areas in Ghana children are not given meat or fish, because it is believed they cause intestinal worms; as a partial substitute, health education emphasizes the value of protein-rich beans and pulses (communicated by Jean Pinder).

# 6

## social
## barriers
## to change

In traditional communities all social interaction is based on well-recognized norms of exchange and reciprocity. Individuals cooperate with other members of their families, with friends and neighbors, and with more distant kin and fictive kin, not because they feel this promotes village welfare, but rather because they recognize that, over time, they benefit in a degree equal to their contribution. Exchange and reciprocity are personalized to a much greater extent than in more complex societies, where the rules governing the same activities are more often stated in the form of laws, work regulations, and the like. In traditional communities every individual recognizes that he has obligations of many kinds and degrees toward other people in the community; when called upon, he feels compelled to help. Conversely, he knows that he has equal expectations, the right to call on the same individuals for similar goods and services. I have described the workings of the exchange system in Tzintzuntzan by means of a model called the *dyadic contract* (Foster 1967a, ch. 11), because these forms of reciprocity follow the pattern of formal and informal agreements between pairs of people rather than between corporate groups. Mutual recognition of a dyadic contract is expressed through continuing exchanges of goods and services in which, as long as the partners value the relationship, an even balance is never struck. Other anthropologists have found the dyadic-contract model to be equally useful in conceptualizing social relations in the communities they have studied.

In addition to informal dyadic ties binding people together, communities are marked by many other forms of social structure like clans or extended families, neighborhood units such as Mediterranean and Latin American barrios, political groupings, and the like. In their collectivity, the norms of authority and the traditional patterns of interpersonal relations between people in a community have a great deal to do with whether certain innovations are possible. In all societies time has sanctioned the social institutions that exist and has validated the ways in which these institutions and the individuals involved articulate with each other. If we are to appreciate the conservatism of peasant communities, it is essential to understand how change upsets the traditional structure of the community at many points, raising fears and apprehensions, perhaps promoting conflict and altering lines of authority. Innovations that threaten the customary ties that bind, or impose on the individual new or undesired social and (or) contractual relationships, are generally viewed with a great deal of suspicion. Aceves, in describing the problems of change in a Castilian village, points out that resistances to a new technology usually are quickly overcome. But "when changes are introduced that depend upon the creation of *new forms of social relationships as a necessary condition* for the change, strong resistance occurs" (Aceves 1971:128).

In the pages that follow we shall consider barriers in social structure under the principal headings of group solidarity, conflict, loci of authority, and characteristics of social structure.

## GROUP SOLIDARITY

In examining peasant and folk communities, one is impressed by the way in which people hold to an ideal of how they ought to behave toward their fellows. This ideal is reflected in a strong sense of mutual obligation within the framework of family and friendship, a general preference for small-group identification, and a willingness to criticize anyone who deviates greatly from these customary norms.

## Mutual obligations within the framework of family, fictive kin, and friendship patterns

Much of the success of peasant society lies in the fact that the obligations and expectations associated with individual roles are social imperatives. They are not optional, at the convenience of a person; they must be recognized and accepted without question. As long as this attitude is held to, a community will have a high degree of integration in spite of factions and interest conflicts. I recall the case of my friend, Vicente Rendón, and his wife, Natividad, in Tzintzuntzan. I had asked them to join me on a trip to Mexico City, 230 miles away. Neither had ever been there, and the event was talked about with much excitement for many days. Then, the night before we were to depart, the father of Vicente's compadre, Salvador Villagomez, died. This posed what to me would have been a real conflict: the obligation to help with the funeral and the keen disappointment of seeing a plum slip through my fingers. After all, Salvador had other compadres who might help. But there was no problem for Vicente and Natividad. Their duty was clear to them, and they fulfilled it without apparent regrets or questions as to whether alternate behavior would have been possible.

This type of reciprocal behavior, whether between members of an extended family, friends, or fictive kin, fulfills a number of functions. In times of food shortages, lack of money, life crises such as death, and in many other situations, economic, spiritual, and physical support is provided. The mutual obligations of these relationships take the place of many of the activities of more highly developed state forms: social security and welfare, an effective police system, cooperative and credit facilities, and the like.

Reciprocal obligations are most effective in maintaining a society when partners have essentially the same access to resources, when their economic well-being is at the same level. Only under these conditions can equality be maintained in the long run. However, reciprocal patterns tend to be incompatible with the trend toward individualization that characterizes urbanization, migration, and industrial or plantation or mine work, for not all villagers make progress at the same rate of speed. Those few who begin to make economic progress find that their relationships are no longer in balance: More is expected of them than they will receive from their partners. Progressive individuals are thus faced with a

cruel dilemma: If they are to enjoy the fruits of their greater initiative and efforts, they must be prepared to disregard many of the
obligations that their societies expect of them or they must expect
to support an ever-increasing number of idle relatives and friends,
with little or no profit to themselves.

Often people solve this dilemma by continuing to accept
the status quo. Macgregor writes of the Sioux:

> *The Indian who amasses a large herd of cattle, builds a*
> *good home, and receives an income of a thousand dollars*
> *or more but does not distribute his wealth on ceremonial*
> *occasions, such as a wedding or funeral in his family, becomes*
> *the subject of severe criticism and ostracism by his relatives*
> *and friends. This is just what happened to many mixed-bloods*
> *who followed the white pattern of accumulating property*
> *and thus lost favor and status with the majority of the people*
> *in their communities. . . .*
>
> *To receive and feed all visitors, especially relatives,*
> *is still an obligation. It is this custom which undermines*
> *economic development of individual families and keeps*
> *them poor. When there were greater food resources to be*
> *obtained by the skill of the hunter and everyone was close to*
> *the same level of wealth, hospitality did not tax the*
> *individual family too heavily. . . .*
>
> *This is not now the case. . . .. hospitality has become a*
> *burden to the few and a strong deterrent to accumulating*
> *material wealth. . . .*
>
> *The man with a regular salary today becomes a target*
> *for his poorer and ne'er-do-well relatives. "He has enough.*
> *Why should he not feed us? He is my relative," is the*
> *prevailing attitude (Macgregor 1946:113–114).*

Brown has compared the common resistance to wage labor
that he found characteristic of the inhabitants of American
Samoa with the resistance of the Hehe people of Tanzania in
East Africa. In both groups a major cause was found to be the
high value placed upon the full performance of social and political obligations, the time-consuming nature of which made
regular labor impossible.

> *In Samoa, social and political obligations are expressed*
> *and emphasized by elaborate ceremonial, almost ritual.*
> *This takes time, both in preparation and actual performance.*

*It also involves a large display and distribution of wealth,
both foodstuffs and less perishable products. This ceremonial
life is highly valued by them, both for the social values it
expresses and as an end in itself, as something dignified and
pleasant. It would be impossible for a people who spend
most of their life working such as a European peasant
or a plantation wage force (Brown 1957:13).*

The Hehe have a less developed ceremonial sense, but a
strong feeling for social responsibility. They must be ready to fulfill
obligations to a wide circle of kindred at such times as crop
planting, betrothals, weddings, funerals, and sickness. Moreover,
every man, to be respected, must assist at the settlement of legal
disputes, an act that, in view of the African fascination with law
and legal procedures, often takes much time. "It is obvious that
their obligations to their kindred and to the community would
both suffer if the bulk of their time were consumed working for
money." Brown concludes that neither the Samoans nor the Hehe
are good subjects for economic development on Western lines
(Brown 1957:13).

In the southern Gilbert Islands the custom of *bubuti* serves
as a similar brake.

*[Bubuti] permits a person to make a formal request for
the goods or services of another. It cannot be refused
without subjecting the person who chooses to do so to the
powerful sanctions of shame. Money has entered this complex
system of more or less continuous interpersonal relations
in a way which prevents individuals from accruing dividends
or re-investing small monetary profits in better goods. For
example, a wage earner in government services on
Tarawa Island is continually subjected to bubuti requests
for money from his relatives in the southern Gilberts.
He desperately needs his wages to live on and should, by all
rational standards, refuse such requests. However, he usually
abides by custom and sends part of his wages to his near
or distant relatives. . . . We thus see that although economic
possibilities have been somewhat enlarged in recent years
they cannot be fully realized as long as the Gilbertese
adjust them to meet the traditional molds of economic
reciprocity with deep roots in a marginally subsistence based
system of economic distribution (Lundsgaarde 1966:25–26).*

Lundsgaarde points out how the apparent resistance of many Gilbertese to capital accumulation can be understood only when one comprehends the traditional sanctions of the society:

> For example, a person with any form of surplus is fair game for a bubuti request and he has been taught from childhood to avoid hoarding by being generous with his goods or services towards others. Self-esteem and prestige can thus curb the desire on the part of any islander to fully accept the economic attitudes of Europeans that encourage competition, maximization of productive potential, or prestige allocation by conspicuous displays of wealth or other such symbols of "success" (Lundsgaarde 1966:34).

In other instances progressive individuals try to cope with the problem by removing themselves as far as possible from their traditional homes. Colson describes the process among the Plateau Tonga of Zambia:

> The wise hawker attempts to trade outside his own neighborhood, for otherwise the begging of his kinsmen and neighbors is likely to reduce his profits to a minimum or he may find himself operating at a loss. . . . Store owners may also find it more profitable to establish their businesses in distant neighborhoods where "business" [i.e. sharp dealings] rather than neighbourliness is the basis of transactions. But this does not protect them from the demands of their own kin, for the location of the store is known and its assets are clearly visible. On one occasion, in conversation with two store owners of Mujika, I argued that one had shown greater wisdom than the other in placing his store at a distant village away from his kinsmen. The men roared with laughter at my innocence, and then one commented: "No man with wealth is able to move far enough to hide himself from his kinsmen" (Colson 1958:82–83).

And, finally, still others attempt compromise, which is rarely satisfactory. Coleman speaks of a process of community disintegration in Nigeria, which is accelerated by seasonal or migrant laborers who return to their villages with new tastes and ideas. When they are in cities, where they are physically isolated from their traditional milieus and where they earn their own living quite independently of village support, they are subject to strong individualizing influences, which create profound tensions in many of them (Coleman 1958:71). On the one hand, the villager is tempted to sever ties with his family in order to live his own life, while on the other hand the pull of family and community is strong, even though the cost is high.

> *Comparatively few succumbed to the temptation of complete severance. The majority sought a delicate compromise, making a gesture toward satisfying the demands of the lineage at home while trying at the same time to pursue their own insatiable quest for a higher living standard in the emergent urban society. As wages and salaries were not large enough to meet African family obligations, the wage-earning Nigerian inevitably faced an agonizing debt, which induced bitterness over his meager wages and the high cost of living. This led to materialism, venality, and an almost pathological obsession regarding money* (Coleman 1958:72).

### Small-group dynamics

A sense of personal identification with small groups seems to be necessary for most people, to provide psychological security and satisfaction in their daily work. This identification may be expressed in the form of a large family, a circle of friends, a cooperative work group, a small village, or any one of many other forms. Part of the importance of the small group lies in reciprocal obligations of the type just discussed, but quite apart from this, the small group seems to provide a gratifying framework within which to work. Often innovations that upset such traditional groups meet with strong resistance, and, again, unless effective small groups can be created, lack of interest and apathy may greet promotional efforts. People, we know, will sometimes forego comfort, convenience, and economic gain in return for more fun out of life. Here, of course, we are dealing with values and value judgments, which are not always rational in an economic sense.

In many of Latin America's villages women gather on the bank of a stream to wash clothes under conditions that are anything but comfortable. But the pleasures of working in the company of others and of conversation and joking compensate for hardship. The Latin American pattern is duplicated in other parts of the world. Dwight D. Eisenhower is reported as the source for an African example.

*In welcoming the American Council on Education's
convention to Washington, the President made a succinct
point with a personal anecdote: "I have never forgotten my
shock, once, when I saw a very modern-looking village
deserted in a far corner of Africa. It had been deserted
because the builders put running water into all the houses.
The women rebelled because there was now taken away from
them their only excuse for social contact with their own
kind, at the village well. I had been guilty of the very great
error of putting into their minds and hearts the same
aspirations that I had. And it simply wasn't so.
(Time, October 19, 1953:26).*

Dr. Deolinda da Costa Martins, who grew up in the famous
old university town of Coimbra in her native Portugal, told me
a similar and even more graphic story. The town lies along the
north bank of the Rio Mondego, rising steeply several hundred
feet to the university, which is situated on a plateau. Here, as is
so often true, social ecology is reflected in topography and the
wealthy families are concentrated near the top, just beneath the
university. In these homes clothing is washed by professional
washerwomen who come in by the day. Two tariffs prevail: a higher
one, if the clothes are washed in the laundry of the owner, a lower
tariff, if the clothing is washed in the Rio Mondego. If the latter,
the woman goes down the hill with a great bundle on her head,
and in the evening she struggles up the hill with the still heavier
wet wash, at considerable cost in time and effort. Yet, in spite
of the added work, most washerwomen are willing to accept
lower pay in order to have the pleasure of being with their friends
and pass a pleasant day in conversation, rather than to be isolated
for a day of hard labor. (communicated by Dr. Deolinda da Costa
Martins).

Problems of the type illustrated by washing habits often can
be solved if they are examined in terms of the values of the
recipient rather than those of the donor group. Then it is clear
that it is the *social* factor, and not the attraction of the riverbank
itself, that is important. If the social demands can be met, the rest
is easy. In San Salvador I have seen public laundries built by engineers who understood the problem. Here washtubs are lined up
in batteries, hot and cold running water flow to each, and there
is plenty of space to hang clothes. Washerwomen pay a few centa-
vos a day, and with far greater comfort than on the riverbank they

accomplish their work and still maintain social ties with their friends.

Recently when Brazilian architects, working as members of a community-development project in the State of Minas Gerais, tried to design an economical "improved" kitchen that gave more usable floor space with less over-all area, thus saving construction costs, they encountered an unexpected social problem. The architects proposed putting the raised cooking hearth in a corner, so good use could be made of wall space and so the central room area would remain uncluttered. But housewives resisted the new design. In Minas Gerais the climate is rather cold, and there are few social contacts for women. So when visitors come, they gather in the kitchen to drink coffee and talk with the housewife as she prepares meals. With the new kitchen design, the housewife had to keep her back to her guests most of the time, an action both rude and socially ungratifying. The architects were forced to accept a larger kitchen in which the raised hearth projected from the wall in such fashion that the housewife could stand behind it and work facing her guests. Economical architectural planning was not synonymous with good social planning (communicated by Arquitecto Bención Tiomny).

The workings of small-group dynamics are revealed in a different context in Spicer's account of the case of "reluctant cotton pickers," in which Japanese of American and Asiatic birth who had been resettled in World War II in the Colorado River Relocation Center were initially antagonistic but subsequently agreeable to the idea of working out of camp to save a local cotton crop. When the plan was first broached, very few evacuees showed interest. Although the reasons involved are many, an important factor was found to be that, to preserve approximate equality of income for all involuntary campers, it had been decided that wages would be put in a community trust fund, for later disbursement as community leaders might determine. But it was clear that the Relocation Authority administrators and the Nisei leaders themselves had overestimated the solidarity of a unit of 9,000 people. Some people thought the only solution lay in individual incentive in the form of wages paid directly to the cotton pickers. The secret, however, was found to lie in considering the barracks "block" of about 300 individuals as the primary social unit. Internees in several blocks suggested that the money earned go into a block trust fund to be used for improvements in the mess hall, for parties, and for other functions with which there was an immediate sense of identification. This was tried and found to be quite successful. Increasing numbers of internees went out to pick cotton, relations with the local Anglo towns improved, and a spirit of competition developed among blocks. Little by little chance aggregates of people were welded into societies. "This case illustrates," says Spicer, "the strength of local group or neighborhood as a social unit focusing the interests and activities of its members. It demonstrates the greater strength of the local group over that collection of local groups which constituted a community" (Spicer 1952:52).

### Public opinion

The unity of a small primary social group seems essential to the prosecution of much directed change. At the same time, it provides, through concentrated public opinion, the means for discouraging the innovator, who potentially is the most useful man in the group. An example, quoted at length, illustrates the tragic results of unenlightened group pressure in a village in India that was subject to the work of the Community Development Programme.

> Hukm Singh, a rangy man in his middle 30's with sloping shoulders, a shaggy black mustache, and a perpetually melancholy look, is one of the best farmers I have met in India.
>
> He also is one of the most prosperous, with 50 acres in south central Uttar Pradesh state where holdings average about five acres. When the Akola community project was launched, the village level worker (VLW) singled out Hukm Singh as a natural village leader. He had land, resources and a high school education. Other villagers looked up to him.
>
> Hukm Singh quickly took up improved seed, artificial fertilizer and other agricultural practices suggested by the VLW.
>
> When the project introduced mustard seed, Hukm Singh volunteered one acre as an experimental plot to compare combinations of 10 varieties and three fertilizers.
>
> With VLW's guidance he converted three acres of land near the irrigation canal into a profitable fruit and vegetable nursery.
>
> Then Hukm Singh went too far. Although he was a Jat, a vegetarian caste, he bought some white leghorn chicks from the project to raise poultry and eggs for tourists visiting the Taj Mahal in nearby Agra.
>
> The other Jats were incensed. The caste council ostracized him. His neighbors refused to pass him the hookah—that is,

*they shunned him in their evening get-togethers over a
community pipe.*

*Hukm Singh held out at first, but eventually he buckled
under the pressure. He gave the birds to his Harijan
(untouchable) sweeper (Hertz 1958).*

## CONFLICT

In folk and peasant communities there is often found an ideal
self-image of village solidarity and agreement that in fact does
not exist. The reader may recall the illustration in Chapter 2 in
which in an Indian village the ideal of family unity is extended to
embrace the entire community, and where conflicts and lack of
cooperation are regarded as shameful. In spite of this not uncom-
mon feeling, we have seen that, in fact, villagers frequently are
suspicious of the motives of their fellows; they fear they will be
outdone by their associates; and they are reluctant to cooperate
with others. These fears, as pointed out, are often reflected in
factional disputes, in which a village divides into mutually antagonistic power blocks. In other situations, where a community is
horizontally stratified economically and socially, some groups will
have special powers, prerogatives, and interests best served if a
status quo is maintained. Conflict problems are here discussed
under the headings or factionalism and vested interests.

### Factionalism

Some kinds of culture change may involve only a few people, or
even a single person. A man may, without being affected by others,
try an improved plow, build a latrine, or learn to read. But most
of today's induced change is based on the assumption that groups
of people will participate. An agricultural extension service is uneconomic if it is directed toward a single farmer; demonstrations
and follow-throughs require that a number of people develop
interest. A single latrine does little to reduce a person's chances
of fly-borne illness, if all his neighbors continue traditional un-
sanitary feces-disposal practices. And literacy efforts, with provision
for reading materials and perhaps village reading rooms, which are
essential to teach adults to read and write, are, like agricultural
extension, prohibitively costly unless group interest is aroused.

Mass programs, then, must be carried out in such fashion
that a significant number of people will wish to participate and

vocal opposition will be minimized. But since societies in transition frequently are plagued by greater-than-normal amounts of factionalism, this often means that if the members of one faction show interest in a new program, the members of another faction immediately declare against it, without logic and without attempting to weigh its true merits. The bitterness and hostility between factions, and the lengths to which people will go to humiliate their rivals, are difficult to believe. Hollnsteiner, in discussing the problems of eliciting cooperation in community-development projects in the Philippines, points out how partisanship is an inevitable consequence of the personalistic alliance system that characterizes that country.

> *Opposing factions will attack almost any project sponsored by their rivals, finding innumerable reasons why it is bad for the community. Though the project may have the noblest aims, such as improving the health of the children, the other side will surely find something in it to criticize. Every resident knows the real reason for the attack: the enmity between the factions involved (Hollnsteiner 1963:190).*

Dube, with his wide experience in rural work, notes that factionalism is also a major problem in community-development programs in India. Village-level workers, the technicians with the most direct and immediate contact with villagers, must make friends with the people they hope to influence. But the very success the worker achieves also jeopardizes his work. According to village norms, friendly relations and friendships presuppose mutual and reciprocal obligations. Hospitality is always returned in some way, and one's friends must be supported. Establishing friendship with a person identifies the outsider with that person's group or faction, and by implication he assumes the villagers' hostility to rival groups and factions. If community-development personnel identify too closely with one faction, their plans will have only partial success.

> *The members of the friendly group would support the officials not because they are convinced about the utility or efficiency of the programs sponsored by them, but simply as their part in the obligations of friendship. On the other hand the members of the hostile group would feel it their duty to reject anything offered by officials identified with their rivals, even if they saw merit in the program (Dube 1957:136).*

Factionalism is not a Philippine nor an Indian monopoly; it appears to be panhuman. R. N. Adams found the same picture in Guatemala. There, a village social worker, a member of a nutritional research program, found shortly after going to work that the people in one section of a village resented her activities more than those in the other, and "as she made more friends in one barrio [section of the community], she became simultaneously less acceptable to the other" (R. N. Adams 1955:442). Adams, an anthropologist, studied the social structure of the village and found that the two halves, the barrios, were different in customs, religion, economic level, and degree of progressiveness. This preexisting schism was unwittingly intensified by the presence of an outsider who, quite by accident, made a majority of her friends in only one of the two barrios.

### Vested interests

Many of the social and economic changes that are being promoted in the world today threaten, or are interpreted as threatening, the security of some groups or individuals. Native medical curers and midwives, for example, often resent modern medical programs because they fear the competition will be injurious to them. Large landlords sometimes feel that too much education for their tenants is undesirable because it will promote unrest and, perhaps, demands for land distribution. Village moneylenders are opposed to low-interest government-managed credit programs, and merchants object to consumer cooperatives. These people normally are individuals of power and prestige, and they are able to exploit their position if they wish to hinder change.

In peasant communities means of disseminating knowledge are few compared to those in urban centers, and rumor is a powerful force in determining action. Untrue rumors are among the most common techniques used by individuals who feel threatened by proposed changes. Particularly where novelty, outsiders, government agents, and free gifts and services are suspect, the word spreads swiftly, no matter how preposterous it may be. Goswami and Roy describe the gullibility of Indian villagers, which, combined with their suspiciousness, makes them an easy target for rumor mongers.

> Vested interests such as the money-lender or the landlord may deliberately start fantastic rumors. Immediately the Bhadson Extension Project was announced, there were rumors that this was the first step to nationalization of land. When a census of persons was taken, there were rumors

*that anyone who gave his age as over sixty-five would be
killed. When an American Point Four expert appeared on
the spot, there were rumors that the villagers would be
driven out and all land used for setting up an American colony.
These rumors made the task of extension workers hard at
the beginning (Goswami and Roy 1953:305).*

In other situations rumors are not necessary. A feudal landlord
can decide whether or not he wishes extension agents or health
workers in his villages. If he decides against such work, individual
decisions on the part of villagers are not involved. Progressive landlords can be and sometimes are powerful forces in promoting
change; in Pakistan I have encountered landlords who paid for
schools, 4-H club supplies, and other civic improvements. But with
greater frequency they, like the midwife and the merchant, are apt
to throw their weight against change.

## LOCI OF AUTHORITY

Where does authority lie in a family? In a neighborhood group or
*barrio?* In a village? How is this authority manifest in decision making, and what are the processes whereby a group decides on a
course of action, or an individual is permitted to take steps he feels
are desirable? The way in which these questions are answered in any
specific situation will have much to do with the receptivity of a
group to suggested change. Within a village, much authority lies in
the family, defined and vested by tradition. Other types of authority
are found in political structure. Authority may be thought of as existing also in the personality of exceptional individuals who, without
formal powers, influence the behavior of others. Authority likewise
exists outside the village; as has been pointed out in Chapter 2, this
is a basic truism with respect to peasant society. Such authority often
overrides local forms and may be a major factor in the growth of a
community.

### Authority within the family

Individual decision making is such a part of American culture that it
is hard to realize that this is not a worldwide cultural pattern. For
example, in middle-class society in the United States the physician's
recommendation that an appendix be removed probably will mean
consultation with the patient's spouse, but it is not the occasion for
a big family council meeting. In many other societies, however, the

locus of authority and traditional decision-making processes are manifest in family structure and are quite different. Public health personnel working in intercultural programs have found that even in small families a patient often is not free to make decisions that are taken for granted in Western countries. Bailey points out how, among the Navaho, the decision to enter a hospital is reached only after a family conference: A woman and her husband alone are not free to exercise their discretion in this regard (Bailey 1948:1419). Margaret Clark found exactly the same situation in the Spanish-speaking Mexican enclave of Sal Si Puedes in San Jose, California: Hospitalization is a grave and serious step, a family and not an individual problem (Clark 1959:231).

Y. S. Kim pointed out to me a similar situation in his country, Korea. If, for example, a young wife is found to have active tuberculosis requiring hospitalization, the physician must first explain the problem to her parents-in-law, who occupy the position of authority in the family; her husband does not have the right to make the decision. Or if a mother finds her child has malaria, she must first ask the permission of her parents-in-law to take the child to a modern health center. If they say no, she is reduced to patronizing an herb doctor (communicated by Dr. Y. S. Kim). Marriott describes an identical example in the Indian village in which he worked. The father and father's brother of a Brahman girl ill with malaria begged the doctor for quinine, and enough was supplied for a full course of treatment. Three days later the doctor discovered that none of it had been used. "An old widowed aunt who ruled the women of that family had voiced objections, and the whole matter of western treatment was dropped (Marriott 1955:243).

Although in most non-Western communities authority is more widely dispersed within the family than it is in our society, there are curious (to us) instances in which parental authority, which we take for granted, is exercised lightly if at all. These situations can be very frustrating for developmental workers. Hazel Blair was surprised to find that among the Eskimo with whom she worked a child's opinion was frequently asked about whether certain things should be done to or for him. "After apparently getting all the necessary consents for immunization I have been told, 'but he doesn't want it.'" (communicated by Dr. Hazel I. Blair). The anthropologist Herbert Phillips, working in a Thai village, encountered the same attitude:

> More than once I have heard parents say that they disliked giving their child a certain medicine "because it offends the baby; he does not like it. It is better not to give it to him

*than to hurt his feelings." Although I do not doubt that
in some instances the parents were annoyed more by the
screams of the child than anything else, I am also convinced
that for the most part they were respecting his basic
individuality and right to choose, even if in this case it was
seemingly to his own detriment—he knows what is best for
himself. Even the incompetent, helpless infant is an independent
soul with his own will and right of determination, and his
dignity must be respected (Phillips 1965:33–34).*

The practical implications of evidence of this type are apparent.
Technical aid experts, in urging specific courses of action that to
Westerners appear to involve only the individual, must in fact find
ways to communicate with, and gain the support of, the individuals
and groups in whom authority is vested. Marriott points out how
lines of power within families affect use of medical treatment in an
Indian village.

*Whatever the treatment may be that is suggested by
a specialist, it will be mediated and enforced, or perhaps
modified or rejected, according to who is most influential in
that particular family. The exploratory clinic in Kishan Garhi
encountered this problem directly when courses of treatment
were thoroughly "sold" to some members of a family but
were later rejected by others who had controlling voices
in the family. Since families in villages of northern India
frequently lack lines of authority that are obvious to
nonmembers, and since the social worlds of men and women
are sharply divided, authoritative communication by the
medical specialist must aim to include all important family
members of both sexes, if it is to be effective (Marriott
1955:251).*

In terms of the way of life of the peoples described in these accounts, these forms of behavior are not irrational; they are logical
and obvious. This is what makes them so difficult to cope with.
Margaret Clark makes clear the conditions that underlie the pattern
of, and reasons for, group decisions in cases dealing with medical
treatment. She points out that in illness as well as in other aspects of
life, people are members of a group of relatives to whom they are
responsible for their behavior and on whom they are dependent for
support and social sanction. "Medical care involves expenditure of

time and energy by the patient's relatives and friends. Money for doctors and medicines comes from the common family purse; many of a sick person's duties are performed during the period of illness by other members of his social group." Ilness obviously is more than a biological disorder; it is a potential social and economic crisis for an entire group of people. Under such circumstances it is not surprising that an individual *cannot even decide unilaterally that he is ill.* Symptoms must be presented to the group, must be described and observed, and consensus must be reached. "In other words, an individual is not socially defined as a sick person until his claim is 'validated' by his associates. Only when relatives and friends accept his condition as an illness can he claim exemption from the performance of his normal daily tasks." Since illness directly affects an entire group, it is only logical that the group should be expected to participate fully in decisions that must be made (Clark 1959: 203, 204).

## Authority in the political structure

In the simplest primitive groups there is no political authority, in the usual sense of the word. The simple functions of government are carried out through the family, or perhaps the lineage or clan. In all peasant and folk societies, on the other hand, we find institutions quite properly labeled "political" in which various types of authority, coupled with the right to make decisions, may be distinguished from authority vested in families. The nature of these political systems has much to do with the receptivity of a community to change. Perhaps the problem can be most simply stated by asking questions: In carrying out technical aid programs, how does the technician identify the leaders with whom he must work? Where does he find the people who have the formal or informal power, the prestige and influence, who will be instrumental in deciding whether work is to go forward? Are the obvious political leaders the best bet? What is the role of religious leaders? Are there informal leaders whose power can be known only after familiarity with the community? Is it best to work through established leaders, whatever their position, in spite of the fact that they are often conservative because of vested interests, or can new leaders be found or created?

There are no simple answers to these questions. We have come to realize that the nature of leadership patterns in a community is one of the most important of all factors influencing cultural change, but we still know much less about the subject than would be desirable. The United States is so organized for activities of all types, with formal and informal, paid and volunteer, leaders, that we assume competent and willing people can always be found to do the job. It sometimes comes as a surprise to Americans to find that in many societies leadership, as an institution, is poorly developed and perhaps inadequate to guide the group decisions that must be made in programs of major change. Often an individual who steps forward to volunteer his services for a new project will be criticized rather than praised for his pains; his neighbors suspect that he sees opportunity for personal gain, at their expense. In some societies modesty and self-effacement are the norms of good conduct; the individual who shows interest in leadership is, at best, guilty of bad manners, and hence subject to public disapproval. The problem is highlighted in a report from a community-development program in southeastern Nigeria, where Colin Hill found that "in the intensely democratic village society, leadership was lacking and, even where the will for voluntary effort existed, there was uncertainty about how to set about making a start" (C. deN. Hill 1957:19). The New England town meeting, so dear to most Americans as exemplifying simple, uncomplicated democracy in small communities, is just not a common pattern in the world, and programs based on the assumption that this kind of decision-making body exists or can be created will usually fail.

The problem of the locus of political authority is illustrated by Dube in an analysis of a community-development program in India. Not illogically—to agents trained in a democratic environment—it was assumed that persons elected to local offices by popular vote were the proper village leaders with whom to work. The community-development workers largely confined their group discussions and individual contacts to this circle and to other obviously important and respected individuals. These leaders responded enthusiastically, and prospects looked bright for successful work.

The basic assumptions about leadership were, however, only partially sound. These people were leaders, but not the only leaders. As is true in many similar situations, they constituted a group with authority to mediate between the village and the outside world, but there were important areas within the community where they had little or no power.

> The village looked to them for guidance in its general
> relationship with the urban areas and the officials; and their
> help was sought in legal matters, in contacting and
> influencing officials, and generally in facing problems that

> *arise out of contact between the village and the outside world.*
> They were not necessarily looked upon as leaders in
> agriculture, nor were they in any sense decision makers in
> many vital matters concerning the individual and his family
> *(Dube 1957:141; emphasis* mine).

In fact, although the villagers were dependent on them for some kinds of leadership, they were, precisely because of this dependence, suspicious of them. Dube concludes that undue emphasis on working with traditional leaders was construed by some villagers as an attempt of the government to maintain the status quo and thus support the domination of the wealthier people. "A closer study of group dynamics in village communities reveals several different levels of leadership, each with somewhat specialized functions." To work effectively in a community, it is desirable to know these several levels and the areas of influence of each. "An excellent village politician," says Dube, "is rarely a model farmer; the latter is generally an obscure and apolitical person 'who minds his own business.' In adopting agricultural practices people are more likely to follow his example rather than that of the local politicians" (Dube 1957:142).

An almost identical situation has been described in a Philippine village in which the government-sponsored community school program of adult education and community development made primary use of schoolteachers as agents of technical change. Teachers occupy a high status in the Philippines because of their education and relatively high salaries, and their leadership in formal schooling matters is unquestioned. But they are not conceded to have universal knowledge; their area of competence is circumscribed. "To be advised on rice growing and animal husbandry by teachers, who did not themselves participate in such activities, was clearly laughable to a sizable proportion of the adult population" (Sibley 1960–1961:211).

## Authority of the unusual individual

Gifted or unusual people, who may or may not occupy formal positions of leadership in a community, often play decisive roles in bringing about change. If they are looked to by their associates for any reason, and if their actions are apt to be imitated by others, they may be thought of as leaders, regardless of their status in their social group. Two classes of people have been singled out by anthropologists as probably the most significant in bringing about change. Barnett believes that the "marginal man," the deviant who is unsatisfied with traditional ways, is most apt to be an innovator (Barnett 1941). Mandelbaum, although not generalizing, gives substantiating evidence in his description of an unusual Kota tribesman (see p. 125). The other point of view is that the prestige-laden individual is more effective as an agent of change. Individuals who, because of social position, wealth, or other reasons, will be followed by others will in fact be more significant in the long run in promoting change.

There is evidence to support both points of view. Barnett thinks of marginal individuals as the "dissident," the "indifferent," the "disaffected," and the "resentful." Such people, he believes, will be more than ordinarily receptive to innovation, for a variety of reasons. Dissenters have never accommodated themselves to specific cultural demands, and the more courageous and independent among them are more apt to rebel and withdraw openly, and to be attracted by new alternatives to traditional forms, than are persons satisfied with customary ways. The indifferent accept new ways because they are not strongly committed to a custom or an ideal of their society. The disaffected are often those who have been exposed to novel ways, and who, within their own cultures, are denied the satisfactions that alien teachings have led them to want. The resentful, obviously, are those who cannot accept the inequalities built into their social systems, if they are in an inferior position compared with others more fortunate (Barnett 1953:378–410).

In primitive and peasant parts of the world bicultural halfbreeds, and natives who have had more than average contact with foreigners, are particular candidates for the categories enumerated. Indian widows, and half-breed orphans and illegitimate children of white men who lived with Yurok, Yakima, and Tsimshian Indian women, are examples of this group.

> [They] had had a taste of the white man's way of life,
> enough to make them unhappy with the Indian way; and when
> they were thrown back to feed upon tribal values, they
> found them tasteless, bitter, or indigestible. Case histories of
> individuals of this sort are highlighted by rebellious and
> delinquent trends, which testify to their suggestibility and to
> their desperate clutching at any new straw which would
> enable them to fight against or relieve them of the necessity of
> conforming to native standards (Barnett 1953:391).

Keesing finds that mixed-bloods are frequently prominent in movements for cultural or political nationalism (Keesing 1941:289).

One of the most interesting examples of this pattern is reported by Mandelbaum. At the time of Mandelbaum's fieldwork in India in the late 1930s, Sulli was the one Kota tribesman in the Nilgiri Hills who could speak English. Through his vigorous personality he was able to bring about a series of significant changes in several villages. His life history showed deviation since childhood: His desire to become educated particularly illustrates the drive of a restless dissident. He had urged his fellow tribesmen to cut their hair, to abandon their traditional calling as musicians for other tribes, to cease eating carrion, and to abolish the menstrual hut.

> *Sulli himself has gone beyond the reforms he advocates.*
> *The prestige value imparted by contact with an ethnographer*
> *and a linguist lent him enough courage not only to cut his*
> *hair but also to tog himself out in shorts, topee, and*
> *stockings in the style of an Englishman. Although his reforms*
> *are savagely opposed by many, some of the younger men*
> *have followed his example in cutting their hair and in*
> *abandoning other tribal practices. Had Sulli been a weaker*
> *personality, he could not have held his followers or himself to*
> *his schedule of acculturation. Much of Kota culture change*
> *is channelled through and directed by a single individual, the*
> *leitmotif of whose personality is deviation from established*
> *tradition (Mandelbaum 1941:26).*

Of course, we are not dealing with an either-or situation. Both the deviant and the prestige-laden individual may be important in bringing about change. In spite of the interesting evidence of Barnett, Keesing, and Mandelbaum, I am inclined to believe that the influence of the deviant individual usually is less important than that of the prestige-laden individual—a wealthy and respected member of a community, a member of the privileged class. If we recognize the enormous force of prestige in driving people to modify habits (see Chapter 8), it is logical to expect this. Comparative evidence also supports this position. R. N. Adams, directing himself specifically to Barnett's position as originally set forth in 1941, challenges the "marginal man" hypothesis and describes the personality and influence of an innovator in a high Andean Peruvian village he studied, a man of wealth and social position (R. N. Adams 1951). Wellin's interesting account of attempts to introduce the boiling of contaminated water in a Peruvian coastal town through health education also suggests that the prestige of powerful people was perhaps the single most important factor in bringing about limited acceptance of this innovation (Wellin 1955).

It should be noted that two distinct but related phenomena are involved here. The dissident or disaffected individual may himself adopt new forms; but whether his example will be followed by others in the community is quite a different matter. Often, I suspect, he is followed only by other dissidents, leaving unaffected the bulk of the population. My guess would be that, as a practical guide to technical aid workers, the really marginal individual is only occasionally a good candidate as an influential local innovator.

## CHARACTERISTICS OF THE
## SOCIAL STRUCTURE

Some aspects of a social structure, and the associated values, either seriously inhibit change, or cause great difficulties for people who adopt change. A poignant case of the latter is reported by Marwick in his discussion of the clash of values between monogamous Christianity and polygynous African religions. The basic differences between the two systems

> set the convert sharply off from his pagan brethren, and sometimes create tensions between him and them. Take this case, for instance. A young couple—call them John and Edith—had recently become Christians when John's maternal uncle died. According to tribal custom John should have married his uncle's widow, Norma. His acceptance of Christianity made this impossible since it would have involved him in a polygamous marriage. The tension that developed as a result of his refusal was described to me by Edith in these words: "Norma used to threaten my husband, and I wasn't happy until the day she died." Before that day arrived, however, two of the children of John and Edith had died; and their deaths had both been attributed to Norma's witchcraft. The implication here is that those who defy the traditional morality do so at their peril (Marwick 1956:491).

In another vein, in countries like India and Pakistan a rigid caste organization slows cultural innovation. The prerogatives associated with each group are jealously guarded, and attempts to infringe upon them by members of lower groups are resented and often repulsed. Under the traditional system, for example, members of certain different castes could not draw water from the same well, attend the same schools, eat together, or otherwise mingle. The types of physical labor one could do were rigorously prescribed, and violation of the rules was condoned neither by members of one's caste nor by members of castes that normally would do such work. Almost the same words could be used to describe the position of Negroes in much of the United States. In other situations, where force is not present, the basic configuration of a society will have much to do with the types of innovation that will occur.

## Caste and class barriers

In stratified societies and bureaucracies people expect to obey or take orders from those in superior positions of authority or power and, in turn, to dictate to those below them. This limits the free interplay of ideas and opinions that is so important in many change situations. Hunter describes a case in which an American university technical assistance team of professors was meeting with a local counterpart group in a foreign country to work out a public administration curriculum for a proposed night school. During discussions a junior member of the American group took issue with the senior member of the U.S. team. The senior foreign professor expressed great surprise that a junior would disagree with his superior's point of view. It turned out that the junior of the two host staff members had never been asked his opinion, nor had he expressed his opinion, on any of the several points that had been discussed (Hunter 1959: 445).

Straus has described a similar situation in Ceylon where the well-developed national agricultural extension service has fallen short of expectations, in part, at least, because of the rigidity of that country's hierarchical, authoritarian administrative system. Extension operates from the top down. The department works primarily through drives to get farmers to grow various crops, and little attempt is made to find out what the farmer feels he wants. The local extension worker finds himself in an almost untenable position. Usually he is of higher caste than the farmers he serves; because of this he (and the farmers) know of no relationship other than that based on subordinate-superordinate patterns. At the same time the agent, since he does a technical job in a society that esteems desk and literary work, occupies a low status in his own bureau. Efficient operations, even with the best technical training in the world, are difficult under such conditions (Straus 1953).

A similar situation was found in the Indian Community Development Programme. In theory the traditional boss-subordinate relationship within the Indian government was expected to yield to a new concept of democratic teamplay. In practice this did not work out. As Dube describes a particular project, the higher officials were unable to view their role other than as that of inspecting officers, and they continued to supervise and appraise the work of their subordinates in the traditional fashion, thus lending a distinct authoritarian tone to their administration. Furthermore, because of their higher status, many resented any views that departed from or even questioned their own policies and work. Although in theory subordinates were expected to express themselves freely, in practice they realized it only exasperated their superiors. Since promotion depended on the goodwill of these officers, the underlings found it desirable to adopt an attitude of compliance and agreement. Consequently, communication was largely one way, and subordinates received and executed orders from their official superiors, without making the suggestions that, because of their experience, they certainly were entitled to make (Dube 1958:90–91).

Role conflict, stemming from traditional caste patterns, also may hinder innovation. Dube describes how a village-level worker of the Community Development Programme was able to persuade villagers of the desirability of constructing compost pits outside the settlement. In general he had the cooperation of people, who for both aesthetic and hygienic reasons felt it best to deposit manure outside the village, and they willingly dug a number of pits. But even though the village council passed resolutions making it obligatory for villagers to use the pits under pain of fine, many were not used. The reason, Dube found, was that tradition dictates that women must clean the house and cattle shed, depositing dung and refuse privately on or near the property. Even women of the highest castes can do this, since they are not publicly visible, but it would be improper for them to be seen carrying such loads through the streets to compost pits on the outskirts of the village. And since this activity is culturally defined as women's work, no man would wish to do so. The few upper caste families that can afford to hire a servant for this work do use the pits, but in general the old, unhygienic practice continues (Dube 1956:22).

## Basic configuration of a society

A single example will illustrate this point. Brown describes how in Samoa three churches have done serious missionary work. The Congregational church has had most success; the Methodist and Catholic churches have had much less, in spite of the fact that the elaborate ceremonial of the latter church undeniably appeals strongly to tribal peoples. Brown believes the similarity in sociopolitical structure between native Samoa and the Congregational church has been the most important factor in producing this situation. Village autonomy has long been characteristic of Samoa; federalization and alliances come into being from time to time, but they are rarely long-lived, so that ultimately power and authority reside in each village.

> *The Congregational Church fitted into this organization because the individual church is autonomous. It lends itself to village separatism and enables the Samoans to identify church and village as one. . . . The more rigid central control of the Methodists has less appeal. As for the Catholics . . . its hierarchical organization is a serious handicap (Brown 1957:14).*

# 7

## psychological
## barriers
## to change

When people are confronted with new opportunities, acceptance or rejection depends not only upon the basic cultural articulation, a favorable pattern of social relations, and economic possibility, but also upon psychological factors. How does the novelty appear to the individual? That is, how does he perceive it? Does he see it in the same light as the technical specialist who presents it to him? Does it convey the same message? Since perception is largely determined by culture, people of different cultures often perceive the same phenomenon in different ways. In the Mexican village of Tequila the Indians are reluctant to call the priest for the last rites for a sick relative, even though they are Catholics. They have noticed that on entirely too many occasions the patient dies shortly after the priest's visit! (Soustelle 1958:166). Similarly, Robert H. Lowie has told how, as a young ethnologist among the Chipewyan Indians on Lake Athabaska, he was unable to measure his subjects for his anthropometric records: The only time an Indian was measured was for a coffin, and if they permitted Lowie to measure them, they knew they would die (Lowe 1959:31).

What the innovator perceives as a decided and obvious advantage may be seen entirely differently by the recipient. It is for this reason that effective communication is so important in programs of directed change. Communication means much more than mastering a language and presenting ideas in simple and clear speech; it means that new ideas and techniques must be presented verbally and visually and conceptually so that the recipient perceives the potential advantages in much the same way as does the technical agent. This does not mean that change comes about only when these conditions are met; sometimes people perceive a program in a very different light from that of the innovator, but they still like it and willingly cooperate. For example, people may build latrines, not because they appreciate environmental sanitation, but because they perceive the prestige that will accrue to them for their progressive ideas, or because they like the technical specialist, value his friendship, and want to please him.

Differential perception and faulty communication may be barriers in situations in which the change agent and the recipient have different expectations about the proper role behavior of the other. When physician and patient meet in a cross-cultural setting, lack of mutual understanding sometimes occurs because each, from previous experience in his own culture, expects a form of behavior from the other that in fact does not occur. Consequently, changes in health practices may be fewer than the physician hopes.

Perception and communication also bear upon the problem of learning. In the final analysis, permanent change in an individual's behavior means learning and relearning. There is increasing evidence that the difficulties attendant upon learning even simple ideas or operations are not fully appreciated by technical specialists, and that unless a series of favorable conditions is present, acceptance of new ways will be much slower than anticipated. A clear, simple talk and a sound demonstration are often insufficient to teach even new behavior forms that the recipient is quite ready to accept.

In this chapter the problem of psychologically based barriers to change is examined under the categories of basic perception, communication, and learning.

## DIFFERENTIAL CROSS-CULTURAL PERCEPTION

### The problem in general

Franz Rosa, while stationed in Tehran as a member of a Point Four mission, once urged his gardener and his wife, recently arrived from a village, to divert themselves by attending the free movie at the United States Information Service. The pair were delighted when they returned. Dr. Rosa asked them what it was about. "Oh," they replied, "it was all about the American dictator. He is a very bad man and very crazy, and he has a little black mustache." The movie was Charlie Chaplin's *Great Dictator*; since he spoke the language of the gardener's employer, they assumed he was the ruler of the country (communicated by Franz Rosa, M.D.).

Similar stories of perceptual misinterpretation are legion. When the linguist-missionaries Benjamin and Adele Elson were first studying the Popoluca language in southern Veracruz, Mexico, they lived in a native house with stick walls and thatched roof. To bring a bit of relief to the interior monotony they placed on their wall a picture of the black and white scotty dogs that advertise the whiskey of the same name. The Indians showed enormous interest in the picture; in fact, their fascination bordered on veneration. Then it dawned on the Elsons that the only pictures the Indians had ever seen were of Christ, the Virgin, and the saints, to whom they light candles and make other simple offerings. Not wishing to create the belief that Americans worship dogs, the Elsons quickly removed the scotties (communicated by Benjamin Elson). During World War II a story, widespread in the Pacific theater, told of a U.S. Navy health officer who hoped to eliminate flies on the island on which he was stationed. He asked the chief to assemble his people, to whom he gave a health lecture, illustrating the horrors of fly-borne diseases with a foot-long model of the common housefly. He believed he had made his point until the chief replied. "I can well understand your preoccupation with flies in America. We have flies here, too, but fortunately they are just little fellows," and he gestured with thumb and forefinger to show their small size and, by implication, lack of menace to health.

Health problems, in cross-cultural perspective, seem to pose a particularly difficult perception problem, if we judge by the number of examples that have been reported. In Egypt, as in much of the developing world, hospitals are perceived as places where people go to die, not to get well; consequently, there is much resistance to hospitalization because the patient perceives it as meaning his family has lost all hope for him. This aversion is expressed in a famous saying about the Kasr El Eini Hospital, the largest in Egypt: "He who enters it will be lost [dead] and he who comes out of it is born [as a new man]" (communicated by Dr. Fawzy Gadalla).

In Mexico, too, hospitals are still perceived by many as a place one goes when all hope has been lost. Recognizing the psychological block caused by the word "hospital," many private institutions have been called by the term *sanatorio*, which comes from *sanar* (to get well), thus emphasizing an expectation of recovery. Recently the Ministry of Health has recognized the psychological implications of the word "hospital," and it now calls its small rural clinics *sanatorios rurales*, while larger establishments are "Health Centers, with Sanatorium." Unfortunately this strategy cannot be prescribed indiscriminately in Latin America. In Costa Rica the first use of the word *sanatorio* was for a tuberculosis sanitarium, which conveys an even more unpleasant idea than a mere hospital. Similarly, in Honduras the word was first used for a mental hospital, so there, too, the word's utility has been lost to other contexts.

In the diffusion of technology, great care must be paid to terminology. The blood bank came from the United States to Latin America, along with the name, which became in Spanish *banco de sangre*. This appears to have been a strategic error, since most Latin Americans, except for the privileged classes, do not look upon a bank as a place to deposit something valuable to be called upon in time of need. Rather, a *banco* tends to be viewed as an instrument of oppression, something dreamed up by the moneyed classes to exploit the common people. Perhaps a more neutral term, such as *almacén* (warehouse), might have been preferable (communicated by Dr. Santiago Renjifo Salcedo, former Minister of Public Health in Colombia).

A high incidence of congenital hip dislocation—1,090 per 100,000 as compared with 3.8 in New York City—afflicts the Navaho Indians. The condition can be treated successfully by nonsurgical means during the first couple of years of life, and surgically during the next several years. But at later ages "freezing" the hip, which eliminates motion, is the only way to prevent probable painful arthritis beginning at the age of 40 or 45. This is the accepted treatment in American culture, and when medical workers discovered the high Navaho incidence, they assumed it was the answer to the problem among Indians. This was not so. The condition was not considered to be really serious, and perhaps it was actually a blessing, since fate, having dealt this minor deformity, was thought less likely to strike a family with greater ill fortune. Among the Navaho, congenital hip is not a barrier to marrying and having children. But, for a Navaho, a frozen hip is a serious problem: He cannot sit comfortably on the ground or on a sheepskin to eat with his family, and he cannot ride horseback. These handicaps more than offset the thought of disability 20 or 30 years in the future, disabilities that may not come anyway. "From the viewpoint of the Navaho . . . the sole contribution of modern medicine to the question of congenital hip disease was to transform something that was no real handicap, and was almost a blessing, into something that represented a very serious handicap indeed" (McDermott et al. 1960:281).

Differential perception has affected agricultural programs also. In a village in India the extension agent encountered difficulty in introducing improved seed to be used for new purposes. There the farmers grew a degenerate local variety of pea that they used as animal fodder. But when the agent introduced a superior edible variety, the people continued to view it as a fodder crop, and being unfamiliar with peas as food they wondered why they should spend more to buy a seed for a crop to be grown for animal consumption. There was resistance to other vegetables as well. People perceived vegetables as a delicacy, an embellishment to their diet, but not as a staple or necessary item. For reasons of novelty and prestige a few people raised small plots of new vegetables, but because of their traditional attitude toward them as food, little effort was made to extend this cultivation (Dube 1956:20–21).

## Perception of the role of government

In the world today a majority of programs of planned change are carried out through government agencies. Yet in many areas, and particularly among the lower classes and rural groups, suspicion of the motives of government and its representatives is widespread. The demands of taxation and military conscription, and other forms of interference that emanate from cities, have taught the villager that the less he has to do with government the better off he will be. An early program of social work in Egypt was hampered by this fact. Getting acquainted with villagers was not easy,

> because social work in its modern form was not known
> even to many well-educated persons in the area, and because
> of the suspicion of the fellaheen, many of whom were
> doubtful about any move to change or improve the conditions
> of their life. Government officials came to uphold the laws,
> to gather taxes, to follow criminals, or to fine them for
> their ignorance and misunderstanding of state laws (King
> 1958:39; my emphasis).

The anthropologist Hamed Ammar has also pointed out how Egyptian villagers consider both government and government officials to be agencies of imposition and control, and hence to be feared. "To keep away from governmental institutions was always the safest policy" (Ammar 1954:81).

Describing a Sicilian village some years ago, Chapman noted the same attitude: "For the ordinary peasant Italy means little more than the Government, a not too friendly entity that imposes taxes, sends husbands to jail and sons to the army, and in the past even scattered cholera in Sicily" (Chapman 1971:155).

When for centuries the presence of a government agent in a village has meant bad luck, it is not hard to understand why villagers are skeptical when an agricultural extension agent, a medical man, or an educator comes and says, "I'm different. I'm your friend. I'm here to help you." Attitudes stemming from the experience of centuries change slowly. Neisser points this out in connection with a community-development and mass education program in Nigeria. In the several villages where work was first done intensively results were good, but when the campaign was spread to more distant villages, she found that "it took several years for the people beyond these experimental villages to overcome their deep-seated suspicion that the government was 'playing a trick' on them and would take away all the new buildings sooner or later" (Neisser 1955:355). Erasmus has described a similar situation among South American villagers: "Generally, farmers are suspicious of government authorities and prefer to let someone else try the new technic before they adopt it" (Erasmus 1954:153).

In India the government attempted to combat endemic goiter in the Himalaya foothills by selling iodized salt below cost and below the price of commercially traded noniodized salt. This step, the government believed, would drive the latter salt off the market and medical goals would be achieved. But the fact that the salt was sponsored by the government convinced the people that it was worthless and quite possibly dangerous. In order to get people to use this salt it was necessary to declare salt a government monopoly (communicated by Dr. G. S. Seghal).

In much of the world the technical specialist's initial handicap stems not only from the fact that he is likely to be a government agent, but also from the simple fact that he is an outsider, an unknown quantity. In Sirkanda, a Himalaya village, anything unfamiliar —a person or a program of change—is regarded with suspicion. "To take a stranger's advice and change accepted practices would be foolhardy" (Berreman 1963:605). Technical aid specialists, with their missionary zeal, sometimes forget how strange they appear to villagers upon first contact and how long it takes for people to be convinced that they are experiencing something new in human relations.

Suspicion of the government agent and the stranger is not limited to rural areas of developing countries. When Margaret Clark began her research among Spanish-speaking people in Sal si Puedes in San Jose, California, and particularly when she took a census, she found much resistance. It took a long time and much hard work before her new friends really believed she was not a government worker in disguise. When Clark made a list of the strangers who were most likely to appear at the door of her respondents the results were startling: truant officer with the news that a child had been playing hooky from school, a deputy sheriff or a representative of the juvenile court informing a mother that her adolescent son was in a scrape with the law, an immigration authority checking on documents, a sanitarian suspecting substandard sanitary facilities, a building inspector wishing to make sure no code was being violated, tax assessors, FBI agents, and many others. All these people, not surprisingly, "are viewed as potential threats to the security of barrio people" (Clark 1959:232). Obviously, a major problem of workers in developmental programs is to establish a role as friend and helper in a society that traditionally has known few or no such roles.

### Perception of gifts

Another aspect of differential perception sometimes not fully appreciated has to do with gifts. In many technical aid programs it has been deemed desirable, because of the obvious poverty of the people, to offer commodities and services without cost. Acceptance of such programs often has been negligible. The reason is that in many situations people interpret a gift as something without value; if something has value, they reason, no one would be so foolish as to give it away. Hence, the mere act of offering something for nothing is interpreted by potential recipients as meaning that the offering is not worth the trouble to collect or use it. But frequently it has been found that if a token price is placed upon services and facilities, people will accept what they otherwise reject..

Anthony Barker, a Scotch medical doctor who with his physician wife went to Zululand to run a small mission hospital, writes of his shock when he learned that the physician he was to replace levied a small fee for her services. He quickly realized, however, that Zulu patients expected to pay something for their treatment, and even that "it is as true in Zululand as elsewhere that prestige is gained by paying heavily for medical advice" (Barker 1959:50). Although his conscience continued to trouble him, Barker concluded that a standard charge of a shilling for each consultation was reasonable.

> *Without this money we should be many hundreds of*
> *pounds poorer every year which would represent a serious*

*hole in the hospital's budget. With it we reduce some of the*
*utterly frivolous demands on our resources and, at the same*
*time, we find a small charge to be a remarkable comfort*
*to the doctor who makes it, while securing to the patient the*
*right to call his doctor a fool (Barker 1959:51).*

When powdered milk was first distributed in Chilean health centers, it was given gratis because of the poverty of mothers who attended maternal and child health clinics. But few mothers would use the milk; they suspected it was of poor quality or downright harmful. A free gift, in their experience, was suspect. Subsequently a token charge was made for milk; the mothers then perceived that in the eyes of clinic personnel the milk had value, and shortly the supply could not keep up with the demand (communicated by Dr. Victoria García).

This same problem arose in the Rockefeller antihookworm program in Ceylon in the early years of this century: The very fact that the treatment was free was suspicious. A Moorish physician attached to the program countered this suspicion among his coreligionists with what is certainly one of the most justifiable of all white lies: He told them that John D. Rockefeller had been very ill with hookworm and had been cured, and out of gratitude to Allah he had given all his money to cure everyone else who had the disease. The account concludes, "May Allah and Mr. Rockefeller forgive this deliberate falsehood, if it should come to their attention" (Philips 1955:291). This story illustrates, in addition to the perception of free goods, a very important point with respect to planned culture-change: The Moslem physician, familiar with the religious beliefs of his patients, was able to answer their doubts and cast his program in terms that fitted the culturally determined expectations of the people as to why a man would want to give something for nothing.

Most of the examples I have encountered suggest the wisdom of conferring value on goods and services by making nominal charges. In one of the hospitals for children in Alexandria, Egypt, the outpatient clinic was badly overcrowded. The administration suspected that a good many mothers brought their children just to have an excuse to visit with other mothers, or to have a reason to get away from their cloistered homes. Since the services had been free, it was decided to collect a small fee in the belief that this would limit visits to those children actually in need of attention. To the astonishment of the hospital administration, the number of patients almost doubled. The mothers believed that now that a charge was being made the services must be much better than when they were free (communicated by Dr. Fawzy Gadalla).

While serving as community-development consultant in Northern Rhodesia (now Zambia) in 1962, I found general agreement among both African and European workers that services and commodities were more appreciated and more widely used when a nominal charge was placed on them. To illustrate, when a four-month course in home economics was first introduced, no charge was made. Because the courses were popular and oversubscribed, it was decided to initiate a fee of about $15, still a very modest price. To the astonishment of everyone, applications doubled! The explanation appears to have been that the courses were now perceived to have real value, and hence to be highly desirable.

### Differential role perception

In every society people learn the behavior that is appropriate to them, and that they may expect from others, in an infinite number of situations in which they find themselves. Differing perceptions of role behavior frequently cause difficulties in intercultural settings, because the members of each group are faced with behavior they do not expect or do not believe to be appropriate to the setting, and in turn they are unsure as to what may be expected of them.

A widespread example is that of the patient-physician interview when the former comes from a background in which the folk medical curer has given the dominant image of what to expect from a healer. Usually such curers, regardless of the part of the world from which they come, differ from scientifically trained physicians in that they do not ask for a case history. Often they divine the nature and cause of the complaint in some mystical or magical fashion. They work in a leisurely fashion and are expected to have unlimited time to be with the patient. Furthermore, they are thought of as friends who are not preoccupied with immediate payment. Since their curing ability sometimes reflects supernatural power, a gift of superior forces, they are not expected to take financial advantage of someone else's need. Instead, at a later time, when the patient is well, he may give a present to the healer in appreciation of his services.

When patients with this image of the role of the curer find themselves in a physician's office, they are often puzzled, offended, and perhaps rebellious. "Where does it hurt?" asks the physician. "Tell me how you feel. How long has this gone on? Have you been afflicted in this way on other occasions?" And so on. The patient perceives this questioning to mean that the physician is not well trained or not very smart. If the folk curer, he reasons, does not need to ask questions, why is it that a physician who claims far greater knowledge must ask them? Gallop illustrates this differential perception with a humorous little story from Portugal:

> At Canecas, on the very outskirts of Lisbon, a doctor was
> called in to see an old woman, and began by asking
> her where she was in pain. É o Senhor que sabe [It's the
> gentleman who knows], was the unexpected reply. It was his
> business to know, she said, not hers to tell him, and she sent
> him about his business. Clearly her previous medical
> advisers had affected universal knowledge (Gallop 1936:73).

A patient whose image of the role of a doctor is that of a leisurely folk curer may be disturbed because the physician seems anxious to hurry him along, to get him out of the office. Perhaps, too, he has been offended because the nurse or receptionist asked about payment. Quite possibly the patient will lapse into silence, and the interview stalls. The physician, on the other hand, is used to treating patients who expect to be asked questions, who realize that the doctor is busy and has little time, and who assume that payment for services rendered is just. Such a physician is puzzled by the behavior of the patient from a different cultural background, and he may become angered and irritated at his mute subject. At best the patient appears stupid, he reasons, and at worst he is downright uncooperative. Culturally determined behavioral expectations—the differential perception of roles—have produced a near-hopeless situation.

Barker describes this situation in his Zululand mission hospital:

> Real difficulties arise from the patients' conception of a
> correct approach to the doctor. In Europe it is accepted
> as the patient's duty to put his doctor in possession of any
> relevant facts about his symptoms, their intensity and duration,
> the history of any previous attacks of the same nature,
> and any other information which the doctor may require to
> guide his examination and arrive at a diagnosis. Among
> African patients such communicativeness is considered to be
> mere weakness, giving away far too much and leaving no
> opportunity for the doctor to demonstrate his skill for which he
> is being paid. Such complaints as are made frequently owe
> more to a sense of the picturesque and a respect for
> tradition than to the literal Anglo-Saxon truth (Barker 1959:65).

On the other hand, in many instances physicians have been so successful in selling the superiority of their techniques that, unwittingly, they have created new but incorrect perceptions in the minds of their patients about the nature of modern medicine. In most parts of the world the wonderworking hypodermic needle has come to be associated with good medical practice. If people go to a physician they expect an injection, and if they do not receive it they suspect that the physician is deceiving them or that he is not a competent medical man. The prescription of oral medicine or of different diet does not meet their expectations for modern medicine.

The belief in the efficacy of the hypodermic is so firmly established in some places that roundabout measures must be taken to give proper treatment. In Nigeria one doctor reports that when internal medicine is indicated, he sometimes has found it necessary to give patients a simultaneous injection of a saline solution. "We attack the illness in two ways," he tells them. "Medicine goes into your stomach through the mouth, and into your body with the injection. When the two meet inside you they react powerfully and cure the disease." This explanation usually satisfies patients.

In the same country, prepenicillin treatment for yaws not infrequently was accompanied by inflammation where the injection was given, because of poor antiseptic measures. This infection came, in the minds of patients, to be a part of the cure, proof that the medicine was in fact doing its job. When the far superior penicillin treatment was introduced, public confidence was lost because there was no accompanying infection. Eventually it became necessary to add an irritant to the penicillin to produce a mild local reaction following the injection. Then the patient's expectation was met; the reaction was proof that the treatment was valid (communicated by Dr. Adeniji Adeniyi-Jones).

### Differing perception of purpose

In a slightly different sense, differing perceptions of the purpose or goals of programs by planners and technicians, on the one hand, and recipient peoples, on the other, may hinder change. There have been occasions when technicians felt that their work was going well and that their programs and goals were understood and approved by the people with whom they worked, and then, for no apparent reason, people appeared to lose interest or seemed unwilling to follow all the way through. Such cases often are due to the expectations of recipient peoples for fewer or different services; when their minimum expectations are met, they are quite happy and see no need to ask for more. To illustrate, I found in Northern Rhodesia in 1962 that a good many Africans asked only a little from "literacy." If they could count and sign their names they could get jobs as watchmen or truck drivers, which meant a considerable rise in the world for them. With these minimal demands met, they had little incentive to spend the several months of hard work necessary to become truly literate, since they could aspire to very little more than they had already achieved.

In a somewhat comparable situation in Haiti, a group of small landowners were provided with irrigation water on a block of about 50 acres. They harvested a much improved rice crop and were delighted with the results. Next an attempt was made to improve the land still further by contour leveling across property boundaries, but this was unsuccessful. "The owners, already satisfied with their increased production, could not see how leveling would result in sufficient further increase to justify the price they would be required to pay (Erasmus 1952:22).

## COMMUNICATION PROBLEMS

In the process of communication someone initiates action in the form of symbols, which usually are verbal or visual or a combination of both. Someone else interprets these symbols in accordance with a culturally determined understanding of what they mean. When both parties are in basic agreement as to the meaning of the symbols, successful communication results. To the extent that people share a common language and culture their communication is made easy: It will be recalled that facilitation of communication is one of the basic functions of culture. People who speak the same language "agree" upon the meanings of the sounds and sequences of the verbal symbols that make up language. They perceive them in the same way, and consequently they are able to understand each other without difficulty. Visual symbols function in the same way. Common gestures such as those for "come," "go," or "wait" are communication devices that are mutually understood by members of the same culture. To most Americans a dancing and singing beer can on a TV screen means, not that someone has perfected a remarkable new can, but that someone wants to tell us that this is good beer.

Members of distinct cultures may make use of the same or similar symbols, but usually the meanings that are attached to these symbols are different. Consequently, when members of one culture are exposed to the symbols of another culture, these symbols are either misinterpreted or not understood at all. For example, the French name "Jean" is recognized by literate persons as a printed symbol for a name. The Frenchman pronounces it (approximately) "John," and thinks of a man. The American pronounces it "Jeen," and thinks of a woman. The same symbol, but with different meanings, results in faulty communication. In Mexico the thumb and forefinger of the hand held closely together and slightly vibrated tells the observer to wait just a minute. To the untutored American, recently arrived in Mexico, this symbol is meaningless.

### Language difficulties

Verbal communication problems that exist between speakers of different languages are apparent to all. What is not always fully appreciated is that serious difficulties may exist between people who presumably speak the same language, for in a large and complex society the total range of linguistic symbols and associated meanings is so great that no one person can master them all. Consequently, the members of subgroups within societies use specialized vocabularies and ways of expressing themselves that facilitate communication among themselves, but that are not fully comprehensible to nonmembers. American teenage slang probably in some measure is unintelligible to most parents. Again, members of professions and trades, such as doctors, lawyers, college professors, carpenters, plumbers, electricians, all use specialized words and expressions. It becomes so much second nature to speak in one's subdialect that it is easy to forget that these professional "shorthands" often are not effective devices to communicate with people who are not members of the group, even though they are members of the same culture and speak the same basic language.

Collins has illustrated this point with a study designed to find out why patients in postpartum clinics in Florida often failed to follow physicians' advice on diet and other health matters. She took as her sample 5 white and 95 black antepartum patients, divided about equally between those from urban and those from rural areas in the northwest part of the state. Fifty of the patients had had at least some high school education, and a number were graduates; the remainder had had various degrees of grade school education. Ninety-seven claimed to be literate, but most said they read every little except for the Bible and comic books. Collins asked the patients if they understood a list of words common in health education, first asking for a definition, and then using each word in a sentence. The results were discouraging: Thirty-eight patients understood 50 percent or fewer of the words, while only

17 understood 80 percent or more. The average was 11.9 words per woman. The words best understood were "margerine" (100 percent, *if* pronounced "marger-*een*", "pregnancy" (99 percent), "starches" (95 percent), "constipation" and "diet" (each 93 percent), and "dried milk" (91 percent). The words least understood were "enriched" (16 percent), "hemoglobin" (17 percent), "rickets" (18 percent), "proteins" (19 percent), and "anemia" (20 percent). Slight changes in terminology considerably increased comprehension: "Egg yaller" was understood by more than "egg yolk," and "powdered" milk by more than "dried" milk. "Well nourished" meant a "nervous" body, "anemia" was usually understood as "enema," and "rickets" frequently was defined as "shaky baby" (Collins 1955). Obviously, a health education program in this part of Florida would stand a much better chance of success if preceded by a careful study of the local dialects.

Medical personnel appear to be among the worst offenders with respect to poor communication. Margaret Clark reports the case of a physician in a maternal- and child-health clinic in a Spanish-speaking community in the United States who gave routine weaning instructions to a young mother with limited command of English: "Apply a tight pectoral binding and restrict your fluid intake." The physician was especially sympathetic to her Spanish-speaking patients and realized they presented special problems. Yet it had never occurred to her that her routine speech habits were actually a function of a highly specialized profession and that a good many educated American women would have to think twice before they understood her (communicated by Dr. Margaret Clark).

Social science research in medical programs in Latin America indicates that poor communication between physicians and nurses, on the one hand, and health center clinic patients, on the other, sometimes prevents maximum efficiency. In 1952 I attended a "mothers' club" meeting in Temuco, Chile, in which an intelligent and well-trained public health nurse talked to about twenty expectant mothers on the problems of pregnancy. Previously a physician had advised them to "walk 3 kilometers a day," and the nurse repeated this instruction. I asked, and received permission, to interject a question: "How far *is* 3 kilometers?" This precipitated a lively argument. Some mothers said 9 blocks, others 27, and a few 81. No agreement was reached. The mothers had learned their lesson, but in the real sense, there had been no communication, because these instructions were meaningless. These women simply were not trained to think in terms of distance as an educated person would. If this important instruction had been phrased as "walk to the plaza and back," or "to the market and then to the church," or to whatever specific places known to them that would give the approximate desired exercise, real communication would have existed.

One day, during my fieldwork in Tzintzuntzan, eight-year-old Roberta Caro came down with mumps, but she refused to stay in bed despite the fact that she was quite ill. I offered to pay her medical expenses if her parents would take her to the doctor in nearby Pátzcuaro, an offer they happily accepted. Then I lectured Roberta on the importance of staying in bed, avoiding other children, and, above all, *of doing exactly what the doctor prescribed*. Roberta agreed to all of this. To my astonishment, the following day I found Roberta playing with her age-mates in the street. I remonstrated with her mother, who replied, "But the doctor told Roberta that she should bathe." Then I understood. In Tzintzuntzan no one bathes who is ill, a taboo that continues until complete recovery. To announce, after an illness, "I have bathed," is to declare publicly that the speaker is completely restored to normal health. The doctor presumably thought he was fighting germs by telling Roberta to bathe, to stay clean, but her mother not surprisingly, in view of the cultural meaning of bathing, interpreted his words to mean, "Since she can be bathed, now she is well." Health education and medical treatment had a consequence entirely different from that intended by the doctor.

### Demonstration dangers

It is in part because of the recognized limitations in linguistic communication that visual symbols, in the form of demonstrations, posters, films, and the like, are so important in developmental programs. Yet even here there are dangers that the meaning of the symbols to the initiator of the action will not be the meaning attached by the observers. In New Zealand a successful health education poster designed to encourage students to brush their teeth showed a whale jumping out of the water in pursuit of a tube of toothpaste. This poster was reproduced for use in Fiji. The response was immediate and overwhelming: Fiji fishermen sent a rush call to New Zealand for large quantities of this wonderful new fish bait! (communicated by Merle S. Farland, R. N.).

As a part of the health program at Cornell University's Hacienda Vicos Peruvian project, an American color film teaching the transmission of typhus by lice was shown to the Indian workers and their families. The film was previewed by health personnel, and they felt it would be understood by the local people, who had been plagued by lice all their lives. Selected members of the Indian community received special instructions about lice and how they transmit disease, in the hope that they could further explain the problem to their fellows. At the showing the physician explained, through an interpreter, the principal facts of transmission.

A week later members of the audience were asked questions to see how effective the health education project had been. It became apparent that the message had not gotten across and that the louse as a vector was not understood. First, the viewers said they had lice but, as in the case of the fly model on the Pacific island, they had never seen giant lice like those shown on the screen so they didn't see the connection. Second, they said they had never seen sick people like those shown in the film, who had a curious and unpleasant white and rosy color. Perhaps, they speculated, this was a disease that afflicted other kinds of people, but again they saw no connection with their own problems. Finally, since they were not familiar with movies, they did not perceive the episodes in the film as a continuum, but rather as a great many odd scenes with no visible relation to each other (communicated by Carlos Alfaro, M.D.).

In this example the health program, while not appreciably furthered, was probably not seriously hampered. On other occasions, unsuccessful demonstrations, or demonstrations that are carelessly carried out, may actually prejudice the viewers. In programs of directed culture-change a skillful demonstration is undeniably one of the most effective communication techniques. In agricultural work, for example, the extension agent will ask for the loan of a small plot of land, which will be planted with improved seed and cultivated in a fashion considered to be superior to traditional ways. If the work is based on careful research, and if the agent knows the proper steps, the demonstration plot should produce a greater yield than adjacent fields cultivated with the old methods. Or the agent may request permission to fertilize a part of a field so that the results may be compared with the unfertilized area.

But if demonstration and other techniques are not skillfully carried out, they may actually set programs back, thereby constituting barriers to subsequent change. Viewers perceive what actually happens and not what ought to happen. The need for workers in change programs is so great that not infrequently imperfectly trained people have been asked to do work for which they were not yet fully prepared. In 1955 I spoke with an East Bengal peasant who complained that the local village worker was incompetent. It turned out that the worker had put too much ammonium sulphate fertilizer on the peasant's bottle-gourd plant and the plant died. In a similar case in Pakistan a number of people vaccinated against smallpox subsequently died of the illness; investigation revealed that the vaccine had lost its potency because of lack of refrigeration. Vaccinators who came the following year found their task almost hopeless. In India in the same year I noted cases in which highly touted "improved" seed had not been thoroughly tested for the local area. It turned out that local soil and climatic conditions were sufficiently different from those of the experimental station that the "improved" seed proved to be inferior to the local variety. Here again the agricultural extension agent subsequently found farmers to be skeptical of his claim that he could help them increase crop yields.

These dangers are particularly great when local workers are under pressure from their superiors to show results for national targets. Dube tells of instances in which village-level workers (VLW) in India were under orders to do things that were clearly unsound. In some cases the seed sent to project headquarters obviously was inferior, and it was apparent that its use would not produce the desired and promised results. In spite of this the village-level workers were under pressure to distribute it. Their misgivings were amply justified, and the harvest was very poor. "In the following year the VLWs found it difficult to sell more seed though this time it was certainly excellent in quality" (Dube 1958:187).

## LEARNING PROBLEMS

The adoption of new ideas and techniques means that people must learn (and ofter, unlearn). The psychology of learning has been well studied and described in recent years, but an awareness of the complexity of the process has not always been translated into action in developmental programs. On one level, Dube has pointed out how Indian village-level workers have complained because they felt inadequately trained for some of the tasks assigned them (Dube 1958:184). Tasks that seem perfectly simple to one who has mastered them may appear very difficult, and perhaps not worth attempting, to people who have not had opportunity to master them. During fieldwork in Tzintzuntzan I have been impressed with the importance of a combination of circumstances favoring the learning of new techniques. Two examples will illustrate.

About 1950 the local tax collector, an outsider, brought the first bicycle to the village. His assistant, a local youth, had as one duty tending the gasoline pump that provided water for the village. The pump was located about a mile away, on a good road. By foot the daily round trip meant an hour or more; by bicycle it could be accomplished in fifteen minutes. The young man quickly learned to ride the tax collector's bicycle and saved money to buy his own.

In a second instance, a village family learned to eat oatmeal for breakfast. A student from CREFAL, UNESCO's Latin American Fundamental Education Training Center in Pátzcuaro, while doing village research and living with the family, asked to have oatmeal for breakfast. Since the dry skim milk, the packaged oats, and salt and sugar were readily available, this request was easily granted. Family members sampled what was left over; at first they disliked it because it seemed slimy. But with repeated samplings they developed a taste that produced a change in their breakfast habits.

In these two instances, the same combination of circumstances was present. First, there was the *continuing presence of a teacher*, over a number of days, to give instruction at the instant is was wanted: Learning was not based on a quick and hurried demonstration or two. Second, the learners had the *opportunity to try the novelty*, to experiment without investing money. Their interest in no way threatened the basic security of their families or diverted funds to untested ends. Third, each had the *opportunity to convince himself of the utility* of the innovation before making a financial investment. They didn't have to make their payments on faith, on the say-so of someone else, but only after they had tried and seen the results. Fourth, in both cases the *cost of the novelties was within the resources of the learners*. If any of the four conditions had been absent, it is much less likely that the innovations would have been adopted.

I suspect that this, or a similar combination of favorable circumstances, will be found in many instances of successful mastery of new techniques, and, to the extent that such circumstances can be created innovation will be promoted.

# 8

## stimulants
## to
## change

In the face of the imposing series of barriers to change described in the preceding chapters, one may wonder how advances can ever be made. Yet change does come, and at an accelerating rate with each passing year. As we have seen, rural peoples are increasingly in touch with urban centers, through radio and television, relatives who have migrated to cities, and improved education in their own villages. People are pragmatic; once convinced that old ways are less desirable than new ways, there are few individuals who will not make major changes in their behavior. With perceived opportunity, and supportive conditions that make realization of success a reasonable hope, social, cultural, and psychological barriers can weaken or dissolve in remarkably short order.

In this chapter I am concerned with two interlocking aspects of the change process: The basic conditions that favor change, and the kinds of knowledge that are helpful to the change agent working in developmental programs. With respect to the former, it is clear that *real* and not *illusory* opportunity must exist. The critical factor here is *how much real choice* the peasant has, given the power and socioeconomic structures that usually characterize his environment. Students of developmental problems generally agree that traditional, semifeudal land tenure patterns, in which peasants are either tenants or owners of marginally productive lands, set absolute limits to the improvements that an individual can expect. Unfortunately, a very high percentage of rural peoples in developing countries live under these conditions. Land redistribution has been a part of the policy of many governments of developing countries, in some cases—Mexico is an example—going back for as much as fifty years. Yet the rate of progress has been distressingly slow, usually inadequate to meet the rising expectations of rural peoples who, thanks to modern media, are growingly aware of how short their end of the stick has been. This awareness, when coupled with skillful leadership, may lead to major changes in the socio-political and economic structures of entire countries, as in the case of Cuba and, more recently, Chile. Or, as in Bolivia, peasants may rise up and forcibly dispossess owners of their estates. Very often in such instances total agricultural production falls, but greater equality of access to resources results.

As an alternate course of action, the urbanization route offers differential rewards in different countries. As we have seen, in countries such as Mexico, Argentina, and Peru, village migrants to cities do surprisingly well in finding jobs and raising their standards of living. This seems due to the relatively high degree of industrialization in the first two countries and even to the more moderate level in the third. But in most African countries industrialization is far less developed, and the new cities seem not to offer rewards similar to those of Latin America. In spite of this differential, the urban flow is equally great in Africa, simply because of the lack of rural opportunity.

In order to study the dynamics of the change process from the point of view of the individual, let us assume, without specifying in detail an optimum political-social-economic environment, that in either country or the city there is reasonable opportunity for improvements in the standards of living for those individuals willing to change traditional behavior. What are the basic conditions that must prevail? At least four points may be noted: (1) The individual must recognize a need and perceive its achievement as possible; that is, it must be a *realistic* need: (2) The individual must have information on how that need can be met; that is, he must know how to go about achieving his goal: (3) He must have access to whatever materials or services the achievement of his goal requires, and at a cost that he can afford. (4) His society must not impose excessive negative sanctions on him for innovating;

that is, the collectivity of barriers discussed in preceding chapters must not be sufficiently strong to dissuade him.

With respect to the strategy of helping people to change, it is important to remember that there are enormous differences in innovative spirit among people within the same community. Obviously it is essential for the change agent to be able to identify likely candidates for his program, whatever it may be, in the shortest possible time. In my work in Tzintzuntzan I have tried to identify the characteristics of the people who, on the basis of information about personal habits, education, political activity, and levels of living, are innovation prone. The first thing that strikes me is that innovation tends to run in families. While an innovator may appear in almost any family, there is a strong tendency for innovative parents to produce innovative children, and so on through grandchildren. This suggests that a certain open-mindedness with respect to new ideas is learned, rather than innate.

A second characteristic of innovators is that they or their ancestors are much more apt than noninnovators to have moved around from one village to another; they are less rooted in Tzintzuntzan. A person who himself comes from another community, or whose immediate ancestors were immigrants, is much more open to new ideas than is the one whose ancestors have always been Tzintzuntzeños. To reduce these data to a principle, we can say that geographical mobility correlates with innovation. Finally, I have found that occupation is strongly correlated with innovative tendencies. Particularly striking is the fact that potters, who constitute more than half of all workers, are measurably more conservative than farmers, fishermen, day laborers, and merchants: They are the least literate, the least traveled, the least political, and the least well housed of the five groups. The explanation for this behavior is, I believe, rooted in the productive process. In order to ensure his narrow economic margin of security the potter must have the highest possible number of saleable pots from each kiln firing. Producing successful batches of pottery, kiln-load after kiln-load, requires meticulous attention to detail and to tried and proven procedures, for at a hundred points in the productive process error or carelessness can lead to disaster. The potter who rigorously follows the techniques that have worked in the past, and who resists the temptation to deviate from usual methods, will be the most successful. This cautiousness which spells relative security in potting, seems to me to reinforce the essential con-

servatism of the villagers, making potters as a class the most reluctant of all people to try new ways (Foster 1967a:295–301).

There is a good deal of evidence to suggest that the conservatism of Tzintzuntzan potters reflects a basic conservatism in
the psychological makeup of potters in other parts of the world.
As a rule-of-thumb guide to community development work, the
implications are clear: Avoid pottery-making villages as initial
targets.

In addition to recognizing the four basic conditions that
facilitate change and recognizing the importance of identifying
innovative people with whom to work, the most successful change
agent knows how to deal with the cultural, social, and psychological
forces that are in opposition to the forces of conservatism. These
forces I call *stimulants*. When in a particular society stimulants are
few and poorly developed, the culture remains static; when they
are numerous and strong, change comes about readily. Speaking of
community development, Jackson describes the picture clearly:

> When a village is faced with a suggestion of change,
> there exists a balance of forces. On one side of the scales are
> those forces which are against change—conservatism,
> apathy, fear and the like; on the other side are the forces for
> change—dissatisfaction with existing conditions, village
> pride, and so on. Successful community development consists
> largely of choosing those projects where the balance is almost
> even, and then trying to lighten the forces against change
> or to increase the factors making for change (Jackson 1956:30).

An understanding of psychological motivations is basic to planned change, yet change depends on much more than the presence
of adequate desire to try something new. Motivations for a general
change may be strong, but unless the innovation fits local cultural,
social, and ideological values, it will have a weak reception. In
the following pages we shall see how these values combine with
motivations to determine, in any specific instance, whether change
does or does not occur.

## MOTIVATIONS TO CHANGE

Motivations are of different types and generality. Some are *culture
bound,* in that their presence or absence is a function of charac-

teristics of particular cultures. For example, religious beliefs in some cultures offer motivations to certain kinds of change, but in other cultures these motivations are weak or lacking. In post-Reformation Protestant Europe the desire to find salvation through personal interpretation of the Scriptures placed a great premium on literacy. A religious motivation— the desire to read the Bible— thus helps explain the world's first mass literacy development. On the other hand, in Catholic Europe, interpretation of the Bible was reserved to the Church, and except for the priesthood the religious motivation for literacy was less pronounced. In 1955 in India I found that the desire to be able to read the sacred scriptures of the Gita, and hence to enjoy a more secure place in afterlife, motivated some people to learn to read, in a fashion reminiscent of early Protestantism (Adams, Foster, and Taylor 1955:56). In most of the modern world, however, it seems likely that—as in ancient Mesopotamia—commercial needs probably will be the single greatest motivating factor in literacy in the long run.

Again, competition and the competitive situation often are strong motivations for change. These are not, however, cultural universals, for in some societies it is considered bad manners to attempt to excel one's fellows. Benedict has pointed out that psychological intelligence tests given to Dakota Indian children were meaningless because of a cultural taboo against giving any answer in the presence of someone who did not know the answer (Benedict 1940:111). And, as earlier noted, in societies marked by limited-good behavior, culture conspires to discourage competition, for the person who might make significant progress as a consequence of successful competition is viewed as the greatest of all threats to the community. In Tzintzuntzan the sanctions against overt competition are so strong that to call an individual "ambitious" is to be highly critical of him. Until very recently, at least, the truly ambitious person tried very hard to conceal the fact that he had competitive urges. Obviously, among both the Dakota Indians and the Tzintzuntzeños, the competitive motivation will be of less importance in stimulating change than it is among native white Americans.

Other kinds of motivations seem to be universal or near-universal in that they cut across all kinds of societies and cultures and are found in some degree almost everywhere. These motivations include such things as the desire for prestige or for economic gain, and the wish to comply with friendship obligations. In the following discussion no formal attempt is made to separate universal and culture-bound motivations, although, when the evidence is good, limitations to the appeal of certain types are pointed out.

Motivations are, of course, of very different intensities. In my experience, two stand out far above all others, and these probably are responsible for the great majority of changes that have occurred and are now occurring. The first is the desire for economic gain; and the second is the attraction of prestige and high status. The two motivations are related, since economic gain usually enhances status, but for analytical purposes it is simpler to consider them separately.

### Desire for economic gain

Earlier chapters gave examples showing how social and cultural factors have caused people to forgo economic gain. Yet in the final analysis these attitudes appear to be a delaying or holding action rather than a definitive barrier. Sooner or later the economic pull seems certain to outweigh other factors. Demonstration techniques, especially in agricultural extension programs, capitalize on this fact. In 1955 in the Philippines I was shown extensive rice fields planted with improved methods. A few successful demonstrations a year earlier had convinced farmers to cultivate hundreds of acres with the new techniques.

In 1962 I noted a similar shift in Northern Rhodesia, where, as in other horticultural (hoe cultivating) societies, women traditionally do the agricultural work. When maize became an important cash crop, and when men realized that their wives alone could not fully exploit this new source of revenue, they were quite happy to take up the cultivation of maize. The same transition took place a few years later with respect to groundnuts (peanuts): When farmers realized that this was a valuable cash crop and that they could augment their incomes by producing for the market, after initial resistance they took over what previously had been a woman's role.

There is convincing evidence that while the taste of a new variety of any food is apt to discourage its adoption as a subsistence crop, when the crop is grown for the market this barrier does not hold. Dube describes how, in the Indian village he studied, there was little enthusiasm for an improved wheat because wives objected to the taste. But improved varieties of sugarcane were quickly adopted by the same farmers because, as cash crops, they brought much higher prices at mills. People remembered the old sugar-cane with nostalgia for its superior taste and allegedly better food value, but, unlike wheat, it was not a staple, and the attraction of the market won out (Dube 1958:133).

Erasmus has analyzed a number of instances in Latin America where attempts were made to induce rural populations to cultivate vegetable gardens for home consumption. He found that vegetable-growing generally was taken up by farmers living near cities or towns, or along reliable communication routes leading to them, but that otherwise, if the farmer found no market, the programs failed after aid was withdrawn. Once farmers grew vegetables for the market, however, they invariably increased their consumption of them so that indirectly, via the pull of the market, nutritional goals were achieved (Erasmus 1954:150).

In Greece the hard-headed and skeptical Macedonian farmers were reluctant to try the new crops and improved methods of culti-vation recommended by Near East Foundation technicians. The big problem was to get the first farmers to try the improved practice. Fortunately most demonstrations were successful, and when these men were seen to have had such a successful year, the trick was turned. "When an improved practice resulted in more cash in hand, the problem of pushing the practice was solved, for then the peasants would clamor for the very instruction about which they had been skeptical before" (Allen 1943:26). A generation later, after World War II and long after the termination of Near East work, a survey was carried out in five villages to determine the long-range effect of this pioneer developmental project. It was found that "practices with obvious direct economic results" had been more completely adopted than practices with less obvious or more indirect results (Beers 1950:281–282).

These examples illustrate two important lessons. First, programs designed to persuade farmers to grow new or improved crops will have a better chance if the produce is destined for the market rather than for home consumption. And, second, once people begin to grow new crops for the market, they are almost certain to begin to consume them at home. Tastes are not absolutely rigid; we all have to learn to like olives.

Mair finds that developments in African land tenure suggest that economic forces ultimately override those of such values as tradition and religion. The modern African's actions increasingly are determined by his perception of economic advantage and not by "an abstract theory of the sacredness of land which inhibits their recognition of its economic potentialities. . . . there are not many recorded cases where someone who was invited to dispose of land for a profitable consideration has invoked such principles as a ground for refusal" (Mair 1957:46). She suggests as a generalization of wide application that

> the conservative force of tradition is never proof against
> the attraction of economic advantage, provided that
> the advantage is sufficient and is clearly recognized. In the
> case of land it is abundantly clear that the emotional and
> religious attitudes towards it which are inculcated by
> native tradition have not prevented the development of a
> commercial attitude (Mair 1957:52).

### Desire for prestige

In all societies there are standards of behavior that are deemed worthy of emulation; that is, if successfully executed they confer prestige and status. The particular acts, or forms of behavior, that confer prestige differ from one society to another. In traditional limited-good societies material symbols of status were conspicuous by their absence, and economic well-being was concealed rather than flaunted. But this did not mean that there were no avenues to the achievement of a respected place in the community. In central Mexican highland communities, and in preindustrial peasant societies in other parts of the world, status was conferred on the man who meticulously, and at great expense over the years, fulfilled a series of ritual obligations to his god and to his community. The achievement-oriented peasant, in affect, exchanged wealth for status; through poverty, brought about by a commitment to the religious values of the community, he achieved the highest prestige the village could confer. In addition to poverty, traditional societies have conferred status on people for chastity, for religious asceticism, for a life devoted to good works, and for many other nonmaterial achievements.

In the contemporary world, however, and even in peasant communities that until a few years ago would have eschewed such ideas, prestige increasingly is sought through acquisition of, or modification in, certain visible symbols such as food, clothing, housing, material items, and speech patterns. Dube points out, for example, how the prestige factor is significant in bringing change to villages of India. Modern Indian, as well as Western, clothes, cigarettes, sun-glasses and tea are popular, both because of their utility and because they possess great prestige value. Moreover, "those who have urban contacts seek to distinguish themselves from the ordinary village folk in their dress, manner of speech, food and recreations. Imitation of urban ways in the last decade and a half has come to be recognized as a distinguishing feature of respectable elements of the village population" (Dube 1955:229).

Hoyt notes a similar development in Africa, where goods and services associated with prestige are in great demand among tribal peoples.

> *Very often the most prestige-carrying goods are something to wear, both because what is worn is conspicuous and because such goods, though not necessarily inexpensive, are relatively so compared with housing, house equipment, and vehicles. . . . Recently there has been a great run in British East Africa for dark spectacles, so much so that an African character wearing dark spectacles is used to typify a vain simpleton, the reason being that dark spectacles are desired for dress wear at social events after the sun has set (Hoyt 1956:14).*

In similar vein, Hogbin has found that in Melanesia the prestige factor largely accounts for the use of corrugated iron roofs on the better homes, which, although more durable, are less comfortable than the old sago-palm thatch (Hogbin 1958:85–86).

In influencing food habits, too, the prestige factor cuts across cultural boundaries, often to the detriment of the health of the individual. Jelliffe has noted that in modern India, mothers of the lower socioeconomic groups in Calcutta increasingly spend money on highly advertised, patented, powdered carbohydrate foods, which have prestige value, rather than on milk and other locally produced foods that represent better nutritional value for the same expenditures. He also points out how two species of fish are raised in the "pond-fish" programs of community-development programs. Both are of equal nutritional value, but one is considered high class, while the other is low class and "snakelike." Relative demand is illustrated by the fact that the latter sells at about half the price of the former (Jelliffe 1957:137).

In the southeastern United States Cussler and De Give found that among rural peoples foods of prestige value, such as those that are purchased rather than home grown, and those thought to be preferred by the upper classes, are desired by lower-class people. The authors conclude that prestige is much more important than health education and other rational programs in determining changes in dietary patterns (Cussler and De Give 1952:36–37, 91).

The same thing holds true for health programs: The prestige factor often outweighs other motivations in leading people to new habits. In Latin America and in India and Pakistan, I have noted that latrines frequently are installed, not because of an appreciation of the desirability of environmental sanitation, but because they add an extra note of elegance to a man's home, and he thereby rises in importance in the eyes of his peers. In Egypt, modern medicine is making some progress because "in serious cases the doctor's help is usually sought besides the traditional cures, and it is becoming a sign of social prestige to bring him into the home of the sick (Ammar 1954:79).

Friedl points out a similar phenomenon in Greece, where the rivalry between village families for prestige is promoting increased use of hospitals. "Hospital care is believed to be the sophisticated, urban way of caring for serious illness and childbirth. Hospitalization therefore can enhance family prestige." This point of view is stimulated by the visits of high-prestige urban relatives who stress the superiority of city doctors over their country counterparts. Hospitalization in nearby cities and towns is the best way to enjoy the attentions of urban physicians. "The result is that childbirth and the care of serious illness at home has come to be associated with low social position and low income, hospital care with high social position (Friedl 1958:26).

Finally, in agriculture, too, the desire for prestige has been found to be a force for innovation. Mosher has described how "using a tractor is, in many regions, a symbol of being modern. Because of this many farmers shift to the use of tractors even when it is uneconomic" (Mosher 1960:13). In the 1920s, says Mosher, this desire for prestige and to be thought modern was an important incentive to mechanization in the United States. In 1962 in Northern Rhodesia I saw many tractors whose use clearly was not economic —in fact, through lack of maintenance they were in some instances inoperable— but they were valued and prominently displayed for the prestige they conferred upon the owner. In 1965 in Nepal I found that the steel moldboard plow was more apt to be acquired for its symbolic value than because the farmer really believed it to be superior for cultivation. But once acquired and tested, a great many of these plows were subsequently put to use, so the initial "irrational" motivation had a happy outcome.

## Competitive situation

People often are spurred to innovation by the motivation of competition. This may be observed between individuals, between groups, and between villages. In traditional rural societies ruled by the image of *limited good,* such competition, as we have seen, has usually been restricted to ritual and religious acts. By underwriting a more costly fiesta than his neighbor, a Latin American peasant declared his allegiance to the community's basic value—not to accumulate wealth—for which he was rewarded with increased prestige among his fellows. Similarly, a village recognized as having an especially large and elaborate fiesta enjoyed prestige beyond that of its neighbors. Among contemporary villagers the old status system based on religion is rapidly breaking down, and anthropologists and folklorists alike note with a degree of nostalgia the erosion of the colorful traditional fiesta pattern. Its place increasingly is taken by competition for visible and material symbols, such as better homes, radios and television, new household gadgets, and the like.

In developmental programs the importance of competition generally has been recognized, and efforts have been made to exploit this motivation in bringing change. Erasmus found that in Latin American 4-H programs results were better when young people worked separate garden plots in competition with each other rather than working the same land cooperatively in such a way that they could not compare results (Erasmus 1954:152). Agricultural and livestock fairs, to which people bring their best produce and animals, often have stimulated interest in programs in these fields. In the Nad-i-Ali area of the Helmand Valley in Afghanistan such fairs, accompanied by judging of orchards, gardens, and fields, with prizes in the form of agricultural tools, have been very successful. Cassel reports that among the Zulu annual garden competitions, accompanied by the establishment of a small market for the sale of surplus produce, constituted a successful technique (Cassel 1955:26).

Tannous believes that in the Near East competition between families or parties within the community may be effectively employed to promote change. He found that in one village two societies tried for several years to excel each other in play production and welfare work. "In most cases it was found that when one family adopted a new agricultural technique successfully, the other families felt challenged to do the same" (Tannous 1944:8).

In Nigeria, when straightforward appeals for progress and development failed, "apathy could often be broken down by arousing a spirit of competition or emulation between neighbouring communities by playing up local rivalries" (C. deN. Hill 1957:21). In the same area, as Jackson points out, the village community attracts the warmest loyalty from its members; "when one village acquires for itself some amenity that its neighbors lack this is a matter for intense pride." This causes envy among neighboring villages, which often strive to keep up. Thus he feels that encouraging competition between villages is a very useful developmental technique (Jackson 1956:35).

The same motivation is found in India. Dube notes that the desire to build up the reputation of one's village is often instrumental in causing acceptance of projects. "A village which built a new school, constructed a public building, or paved all its lanes could earn a reputation in the vicinity for being 'progressive.' Competition between individuals, families, kin-groups, castes, and villages was also at the back of many programmes" (Dube 1958:83).

But competition is not a sure incentive to greater effort. Dube describes a cattle show in which substantial and useful prizes created interest and in which the judging was entirely fair. Still many people were disappointed.

> Some felt that all villages entering cattle in the show should have been rewarded in some way or the other for their co-operation: a village would not lose face if even one of its residents got a prize. In one case it was found that the leaders of the village which had secured a large number of prizes were half apologetic, for their very success was viewed by others as a mark of selfishness (Dube 1958:117).

The same results were noted in crop competitions. Winners were not stimulated to maintain their lead the following year. In one instance a winner said he did not want to be selfish and that others were now entitled to win. Dube found little evidence that anyone adopted new practices or took extra care of his crops in order to win a prize.

Competition can also be dangerous in that it leads people to spend money for prestige symbols that they do not really need. In rural Greece, Friedl found that urban-inspired competition for prestige "often encourages spending beyond the boundaries of economic wisdom, quite on the 'keeping up with the Joneses' pattern." For example, clothing that is not suited to the country and that wears out rapidly is sometimes purchased in preference to garments better suited to country tasks. On the other hand, the competitive urge is not all bad, she finds, for it has produced such improvements as sanitary latrines and increased willingness to accept hospitalization (Friedl 1959:35).

In newly developing nations, the competitive motivation seems ultimately to rise to the national level, where it takes the familiar form of nationalism. The nationalistic spur often is important in bringing about change. Neisser describes how the desire "to take a feather out of the white man's cap" by learning to read and write was an important aid in educational work in Nigeria (Neisser 1955: 355). Writing about the same country, Jackson speaks of the "spiritual discontent" stemming from the African's belief that Europeans regard him as inferior—a feeling that motivates many people to excel (Jackson 1956:36).

A counscious appeal to nationalistic feelings has promoted change in many situations. Tannous tells how in the Near East the evoking of the "glorious past" stimulates the *fellahin* to improve agricultural practices. He found that showing lantern slides of Arab monuments still standing in the Middle East, lecturing on the Arab contribution to world civilization, and giving a laudatory description of their country's ancient successful agricultural practices contributed significantly to extension work (Tannous 1944:8). Among the Zulu, where diet had deteriorated over the years, the aims of a comprehensive health program were furthered by pointing out to people the values of the old diet, and by stimulating their pride in past ways and their desire to maintain former standards (Cassel 1955:24).

In India, the speeches of community-development leaders often contrast the country's present problems with its high position in antiquity, and villagers are asked to help restore this former glory through hard work. If the United States and Russia, they are asked, could build up their national prosperity from lowly beginnings, why could India not do the same? (Dube 1958:109–110)

## Obligations of friendship

In much of the world, friendship patterns are more carefully worked out and balanced than they are in the United States. Disinterested friendship often plays small part in the daily relations between people. The mere fact of the recognition of the friendship means that the partners assume mutual obligations and expect reciprocal favors in return. Pitt-Rivers's classic account of friendship in the Andalusian village of Alcalá points up the nature of relationships that will be encountered when a technician—or an anthropologist for that matter—enters a folk or peasant community and establishes contact with local people. In such cases, says Pitt-Rivers, friendship is based on a "spirit of contract." Friendship is the free association with a person of one's choice: It implies mutual liking, but it also involves mutual service. "To enter into friendship with someone means putting oneself in a state of obligation. This obligation obliges one to meet his request, even though it involves a sacrifice on one's own part. One must not, if one can help it, say 'no' to a friend." In Alcalá, a friendship is established through a favor that expresses one's *simpatía*, or mutual liking for a person. If the favor is accepted, the bond is established, and one is then entitled to expect the return of favors (Pitt-Rivers 1954:138, 139).

If friendship is looked at in this light, rather than in the more casual American way, many of the reasons for cooperation on the part of villagers will be understood. In 1955 I found, in asking villagers in India why they had accepted certain community-development projects, that the frequent answer was, "To please the village-level worker." The villagers were not necessarily convinced of the desirability or utility of the action, but they had come to feel indebted to the VLW, and the obligations of friendship, as they understood them, required them to help their new friend. Similarly, Dube discusses the instructions of village-level workers to "fraternize" with villagers. "Fraternization involved establishing friendly relations, and friendships, according to village norms, involve a series of mutual and reciprocal obligations. Hospitality must be returned in one form or the other, and one must support one's friend." Factional dangers were present in the villages Dube studied. "The members of the friendly group would support the officials not because they are convinced about the utility or efficiency of the programs sponsored by them, but simply as their part in the obligations of friendship" (Dube 1957:136).

In the UNESCO-sponsored fundamental education project (CREFAL) in Pátzcuaro, Mexico, a great deal of the cooperation encountered in one village can be explained by the fact that one of the specialists previously had worked there as an anthropologist, and the friendship ties that had developed between him and many villagers obligated them to participate in a number of activities they would have rejected if they had been left to their own judgment. The same fact of friendship was the principal motivation for seven of the eleven Peruvian housewives in the village of Los Molinos who heeded the advice of Nelida, the health educator, and decided to boil drinking water regularly (Wellin 1955:83).

### Play motivation

The significance of the play motif in the processes of invention and discovery is well known. Most scientists—frequently as a form of relaxation—like to play with their data, to arrange facts in new combinations, or to pursue a line of investigation for no other reason than that it is fun. Sometimes important new ideas emerge from this play process. Play also is an important factor in culture change, but unfortunately it has not been well documented. Perhaps it is best seen in the United States, where we often try a new tool or gadget just for the fun of it, out of curiosity .But this love of play does not seem to be culture bound—it can be expected to crop up on any cultural level. The Trumai Indians of central Brazil are among the most primitive of all peoples, yet when they were visited by the anthropologist Buell Quain in the late 1930s, they were tremendously interested in his belonigings and made continual attempts to acquire them. "Adults as well as children whined in order to obtain objects from him. This attitude was manifested towards useful objects such as knives and axes, but they also pleaded and begged for articles having only curiosity value to them. . . . to be curious about an object was to want it" (Murphy and Quain 1955:41).

Among the South African Zulu, Cassel found that the curiosity of the women about how a fetus was nourished *in utero* led to much discussion in prenatal clinics, thus facilitating the use of posters and models showing the functions of the placenta and umbilical cord (Cassel 1955:24–25). Dube reports that several items in the community-development program he studied, such as new agricultural implements and improved techniques, were tried out of curiosity. Although some innovations were turned down because people found them unfamiliar, others were enthusiastically adopted because of their novelty (Dube 1958:84). And in Greece, Allen describes how improvements in farming not infrequently appealed to the Macedonian farmer as fads (Allen 1943:248).

I encountered an unusual example of the effect of the play motif in East Bengal in 1955. In one village to which I was taken nearly 80 percent of the families had built bore-hole latrines, all within the space of several weeks. I asked the American technician how he explained this. He believed that it was because the health lectures given by his Bengali associates had been carefully worked out, and the people had listened with care and been convinced of the desirability of building latrines. Since such success had never before occurred in any environmental sanitation program with which I was familiar, it seemed likely other factors were present. Investigation proved this to be true. In this part of East Bengal there is a thick covering of rich alluvial soil, which permits the use of an auger for drilling the latrine pit. Four men can bore through as much as 20 feet of this soil in an hour or so, and the results are little short of miraculous. It turned out that the villagers were enchanted with this marvelous new tool, and all wanted to try their hand at it. They felt that the concrete perforated slab they had to buy to cap the hole was a small price to pay to enjoy an hour or two with this wonderful new toy. Competition between groups of men was informally organized, and records were set and broken in rapid succession. For several weeks this was undoubtedly the happiest village in the country. And, at the end of the time, a good job of environmental sanitation had been done—but not for the reason the health team thought!

## Religious appeal

Examples have already been given of how a religious motivation was instrumental in two areas in creating a demand for reading. Sometimes conscious use can be made of a religious motivation if a developmental worker is familiar with the religious beliefs and sacred writings of the people among whom he works. In several villages of the Etawah Pilot Project in community development in India, acacia trees were planted on an acre of wasteland near the villages and carefully watered and tended so that the young trees were not destroyed by goats. How was this unusual cooperative undertaking achieved? "The land was solemnly dedicated by each village as a 'Krishna grove' (*Krishnaban*), so that there was real emotional attachment to it as a sacred undertaking, sacred to Lord Krishna." The villagers were determined that they would not "let Krishna down," as had happened in other communities (Mayer and Associates 1959: 215–216).

The same technique was used—by lucky accident—in Future Farmers of Greece clubs organized as a part of the Near East Foundation program in that country. In some villages as many as 9,000 treelets were successfully planted. Although it was not a conscious part of the strategy of change, the village priest was always called upon to participate. "In simple, Orthodox ritual [he] blessed the willowy saplings before the lads put them into the ground, while the villagers looked on approvingly" (Allen 1943:230). This religious sanction was one of several conditions that favored the work.

In working with the *fellahin* of the Middle East good results often can be attained by the use of citations from the Koran, thus adding the supreme religious validation to the lesser authority of the technician. Expressions such as "Allah hath said in His exalted book" and "All those who love the prophet" and "He who is willing among you to honor his religion and Allah's book" seldom fail to arouse active response. In addition, the Koran contains a wealth of statements pertinent to literacy, health, and improvements in agriculture, which can be cited to clinch the truthfulness of an argument for innovation (Tannous 1944:8). Examples of such statements, which can be used in health and developmental programs, are given by Gadalla: for personal hygiene, "Cleanliness is a part of Islam"; to encourage dental hygiene, "If it were not hard upon my people I would have asked them to brush their teeth before each prayer" (That is, five times a day); and to stimulate cooperation, "A Moslem to a Moslem are as the walls of a building supporting one another." Gadalla emphasizes the importance, in a Moslem country, of the worker's introducing his task with the phrase "In the name of Allah." This shows that the technician is a good, religious man. When in practice in Egypt, Gadalla always used this phrase before beginning an operation or an examination, and before giving an injection or dressing a wound (communicated by Fawzy Gadalla, M.D.).

### THE PROBLEM OF "FIT"

In studying examples of social and cultural change one encounters a series of factors that have contributed significantly to the introduction of new forms of behavior and that, although motivation obviously is involved, can better be illustrated in terms of "fit." These factors involve such things as cultural values, social forms, motor patterns, and economic reality.

### Social forms and values

An innovation, to be successful, requires among other things a supporting social structure onto which it can be grafted. This simply means that in all societies traditional institutions have recognized roles; if new forms can be integrated or associated with these traditional roles, they have a better chance of being accepted than if there is nothing to tie to. A scientifically trained physician ,whatever his problems may be in working with people whose previous experience has been with folk medical curers, has at least one point in his favor; his role—that of a person who tries to make people well—is known and appreciated. The health educator, on the other hand, labors under the handicap that traditional societies make no provisions for people with his function. Jackson illustrates this point clearly with an example from Nigeria. Here, as in much of the world, maintenance of new facilities is a major problem. When better and more sanitary village markets were constructed, the improvement was kept up largely because there was a preexisting system for cleaning the market after the day's sales; the social form necessary for caring for the new facility was there, and no new organization had to be set up. But when a new project such as a reading room was completed—a project that had no parallel in village life—it proved almost impossible to create a new organization to care for it (Jackson 1956:101).

In the Lake Pátzcuaro region of Mexico the UNESCO-sponsored CREFAL fundamental education program has been successful in introducing chicken farming in a number of villages. The local people have kept chickens for centuries, and the White Leghorns have generally been recognized as superior birds. Here the fit between values, social forms, and economic possibilities has been good, and the program has been gratifying in results. Contrast this with a similar case in China several years ago: White Leghorns, the best bird for the area from the standpoint of the poultryman, were introduced as in Mexico. But the innovator was unaware of a tabu against raising and eating white birds; consequently little progress was made (Yang Hsin-Pao 1949:18). It seems likely that a bird such as the Rhode Island Red might have been introduced successfully, since it would offer no fundamental conflict with local values; the fit would be adequate.

Similarly, in parts of Mexico pigeons and doves are believed to bring bad luck in the form of illness, death, or desertion by one's spouse. Developmental programs that have tried to introduce these birds in such areas have not met with success (Kelly 1958:206). But emphasis on chicken farming might well meet with the success accorded it in Pátzcuaro.

### Motor patterns

The motor patterns of a group fit the tools and activities that characterize daily life. Developmental programs usually involve new material items and new customs whose fit to urban or Western culture may be perfect, but that are not in perfect harmony with local ways. We have already seen examples of barriers due to differences in motor patterns. To the extent that new tools and techniques can be adapted to preexisting ways of using the body, the possibilities for successful introduction are multiplied. Often native peoples display great ingenuity in reworking the products of the machine age.

Firth tells how, in Tikopia, an outlying Polynesian island in Melanesia, steel tools replaced the aboriginal giant clamshell blades: When the Tikopians acquired the wood plane, instead of using it to smooth timber, they removed the blade and hafted it to an adze. The adze is the traditional woodworking tool, and they found the new blade a significant improvement over the clam shell (Firth 1956:94).

Many years ago Thompson described how Mexican farmers, who customarily use an ancient Mediterranean-type single-handled "scratch" plow, modified American two-handled plows by removing one handle. The tool was easier to modify than the motor pattern. He also described how, to the astonishment of American and British railway construction gang foreman, Mexican peons, when furnished with modern wheelbarrows, removed the wheel and lifted the barrow proper to their backs, supporting and carrying it with a forehead tumpline! The Mexican, since the time of the building of great prehistoric pyramids, has moved earth by carrying it in a basket supported on his back in this fashion. Even today the tourist in Mexico City may see earth from the foundations of new skyscrapers being carried out of the ground in this same way. The knack of handling a wheelbarrow is not something an adult picks up easily, so the willing railway workers, not wishing to displease their masters by rejecting their help, solved the problem by changing the tool to conform to their motor patterns (Thompson 1922:52).

The lesson of the plow has been learned by a great many agricultural experts in United Nations and Point Four programs. Often —but by no means always—lightweight one-handed moldboard plows are being introduced into those parts of the world where the traditional plow has but one handle, and in such instances a major factor of resistance is overcome. The technician on a foreign assignment who is willing to study local motor patterns and is able to find ways to modify new implements that he is introducing, sometimes finds that this simple operation spells the difference between success and failure. This is not a new discovery. Two hundred years ago British rifle manufacturers who sold to the Iroquois and other Indians of eastern North America discovered that the stocks of their products were wearing away at the top, and that customer resistance followed. In the thick underbrush of the eastern woodlands Indians customarily dragged their long bows after them, one tip riding lightly on the ground, to avoid entangling them in thickets. For the same reason they dragged rifles along the ground, but because of their greater weight the wooden stocks wore away. An ingenious manufacturer solved the problem by designing a special iron butt plate that curved over and protected the part of the stock that came into contact with the ground; presumably, he thereafter had satisfied customers (communicated by W. N. Fenton).

## Sequence

The particular instant at which an innovation appears in a change situation will have much to do with its acceptance or rejection. As has been pointed out, the success of an innovation depends in large measure on the supporting circumstances that may exist, as well as on the recognition by people of the need for the new thing. But in a period of rapid change, the number and strengths of the factors that improve the chances for acceptance of a specific item will fluctuate over the years. If the innovation appears at that point in time when the supporting factors "peak," the chances for acceptance are excellent. If it appears in the "trough," rejection is likely. The skilled change agent is the one who can recognize when the time is ripe for the next step in his program and who has the patience to hold back until this time. However desirable an innovation may seem, there is no point in trying to force its acceptance at a time when all change indices say no, not yet.

Literacy and adult education generally are thought to be basic desiderata in community-development programs. Yet more often than not such activities have met with little success. In part the reason has been that they have been presented before people were ready for them. Richardson tells how the village reading rooms of the Near East Foundation's Greek program were failures initially. They simply had no meaning in terms of village life. But after the Future Farmers of Greece clubs were organized, with the stimulation of boyish interest in better husbandry practices, the ability to read and the ready access to sources of knowledge placed a premium on reading rooms (Richardson 1945:22).

Usually the desire to read and write comes late in the development of peasant society. Villagers do not look upon literacy as an abstract thing that is good per se. It is something that takes time and hard work, and when it is achieved it has no meaning for most people. Berreman tells how, in a Himalayan foothill village, attitudes toward government school programs generally are not favorable. Such attitudes are due in considerable measure to the apparent absence of tangible benefits accruing to those people who have been educated in the past (Berreman 1963:335).

Buitrón illustrates this point with an example from Chimborazo, Ecuador, where he arrived three months after a number of farmers had received certificates of literacy following an adult education campaign. When he asked them to write their names, they were unable to do so; in three months they had forgotten even this elemental art *because they had no need to read or write*. For this reason, writes Buitrón, before teaching anything, we must be sure the new knowledge will fill a real need felt by the people; if this is not the case, we must first create the need.

The group just described consists of landless hacienda wage employees. Buitrón contrasts them with another agricultural group, which also weaves and sells cloth. The members of this group, in order to sell, must know Spanish as well as Indian languages, must learn schedules of buses and trains, read house numbers, learn accurately the value of money, and be able to make simple mathematical calculations in order to make change. "Among these people there is great need of literacy. They themselves understand and recognize this need, and without outside advice, send their children to school." Bruitrón concludes that "those who believe that to eradicate illiteracy it is sufficient to augment the number of teachers, better their preparation, and create more schools, are completely mistaken" (Buitrón 1960:169–171).

Jackson describes an illuminating example in Nigeria. During World War II an adult literacy campaign was very successful, but interest nearly vanished at the end of hostilities. It was not for the sheer joy of intellectual mastery that people had wanted to read and write; they had wanted to communicate with their young men who were away as soldiers. When the soldiers returned, there was much less felt need for literacy (Jackson 1956:21). In similar vein, in Ghana in 1962 I was told by community-development workers that the strongest literacy motivation is found in those families with men working away from home in mines or cities. Family members remaining at home feel secretive about the contents of letters from absent relatives; they feel that should they ask a literate friend to read the letters for them, the entire village will soon know how their kinsmen are doing. And they are reluctant to have this known—particularly if economic success is reported.

As in the case described above, literacy generally becomes important only when villagers begin to see that those persons who are literate have extra advantages, or when they come to feel they are less likely to be cheated by city people if they can read. As standards of living are raised and there is increased travel and contact with the remainder of the nation, conditions become increasingly favorable for presenting adult education programs to rural peoples. But until such time arrives, slight success is the usual rule.

A similar pattern of recognized need is apparent in health programs. Public health work is based on the philosophy of prevention of the conditions that lead to illness, rather than on curing the sick. Yet the evidence indicates that in those parts of the world where curative facilities are not readily available, preventive programs have little meaning to people. The immediate needs of the ill must be met before they are interested in immunization and environmental sanitation, the importance of which, from their point of view, is pretty hard to see anyway. Experience shows that the prestige won by curative medicine is one of the most effective ways to sell preventive programs. At the right time in the sequence of a medical program, people will take on faith things that will not attract them at an earlier period.

## Timing

The problems of sequence are long range. But another aspect of the likelihood of acceptance of innovation can be thought of as a function of the right time within the yearly cycle. This is particularly true when people's pocketbooks are involved. In the average rural community, for example, the amount of cash people have fluctuates widely during the year. At harvest time they are relatively prosperous. This is the traditional time for weddings and often for fiestas that involve big expenditures. It is a time when new clothing and household needs are looked after. If new material items or practices come to the attention of villagers at this time, they are much more apt to spend money on them than they would be a few months earlier, when the new harvest has not come in and the resources of the preceding year are nearly exhausted.

Batten puts it well:

> The government officer must not only be prepared to give
> his time, but also to give it at the right time of the year.
> People are much busier on their farms at some seasons than at

> *others, and they can take on project work more successfully*
> *in the slack season when they have the most leisure time.*
> *The project which is stimulated too late, so that it is left*
> *half-done when the busy season comes round again, may never*
> *be completed, and may prejudice the people against*
> *attempting other projects in the future (Batten 1957:55).*

He gives as example the cocoa-growing areas of Ghana where many people leave their villages seasonally to work on distant farms. Here, he says, it is obvious that correct timing is particularly important (Batten 1957:55).

### Middle-class receptivity

What is the socioeconomic position of people who seem to adopt new practices most easily? Well-to-do people, obviously, have the means to acquire many innovations inaccessible to their less well-off neighbors. When it is a question of items of material culture, this group is often the most receptive. But frequently these people are basically conservative. They are content with their position, and if major changes in the way of life of their community occur, there is no certainty that they will continue to enjoy their advantage. Further, such people often have a deeper commitment to dominant local values than do less fortunate people. The reluctance of superior farmers in India to buy improved seed, because of fear of loss of face, will be recalled (Chapter 5).

The poorest people sometimes seem like good candidates for change. They have so little to lose, economically and socially, it is argued, that they are taking no great risk in adopting the new. Many developmental programs in such fields as agriculture and health have been directed toward this target group, both because of the philosophy of dire need and because of the belief that the people will be receptive. But again experience shows that the lower socioeconomic groups are usually the poorest candidates for change. Batten points out why poor people, who would seem to have most to gain through judicious innovation, are usually the most reluctant to adopt new ways:

> *This is because they of all people can least afford to take*
> *risks. They have no reserves to tide them over failure.*
> *They know that they can just make a living by doing as they do,*
> *and they need to be very sure before they do anything*

*differently. After all, it is they, not the worker, who will suffer*
*if the worker suggests the wrong thing (Batten 1957:10).*

In my experience the most receptive people are those who are at neither the top nor the bottom of the local socioeconomic scale. They have enough so that they can gamble with limited experiments without unduly threatening their well-being, but their position is not so secure but that the attraction of greater income, as well as the possibility of satisfying other felt needs, is a strong motivation to action. Usually they do not represent vested interests that may be threatened by major innovations nor, on the other hand, are they sunk in apathy, believing that no real change is possible.

### Authority

The extent to which authority—or, to use a happy euphemism I have heard in India, "executive methods"—should be used in programs of planned change is debatable. Viewed as a social phenomenon that is a part of the dynamics of change, however, there is no doubt that it is an important factor. Actually, at least two kinds of authority can be recognized as influencing receptivity to innovation. The first is that in which a respected leader vouches for, or places his stamp of approval on, a proposed change. He is not forcing people to change; rather, he is using the authority inherent in his office to reassure them that it is, in fact, safe and desirable. Speaking of the Kabyle of Algeria, Bourdieu writes:

*In many Kabyle villages, the old-fashioned teacher . . . was*
*able, thanks to his prestige and moral authority, to introduce*
*many innovations in all areas of life.* Confidence in persons
proposing the transformation engages much greater following
than the reasons invoked to justify it *(Bourdieu 1963:67;*
*my emphasis).*

Most Americans feel that there are times when the good of the majority requires the use of authority over the will and beliefs of the minority. Compulsory vaccination and chlorination of water have brought significant health improvements, over the protests of dissident minorities. In other cases the initial use of authority has meant that people have voluntarily carried on after learning the advantages of a new custom, or at least after coming to the conclusion that they are not being harmed. Dube, for example, describes how in a post-

Independence Indian village a government revaccination program was accepted without resistance, even though people were surprised to learn that their initial immunization was not good for life. Under British rule vaccination had been compulsory, people were used to it, they did not fear it, and although there was skepticism as to its value, people generally were cooperative (Dube 1958:74).

The introduction of the New World potato into Europe affords another example of how the initial use of authority may work to the benefit of all concerned. When the potato was first brought to Europe its acceptance was not spontaneous and unopposed. It was rumored that it poisoned the soil, that it caused diarrhea, and that it was otherwise harmful. Benjamin Thompson (Count Rumford), a royalist who left his native Massachusetts in 1776, became military advisor to the Duke of Bavaria. Noting the frequent crop failures of those days, he felt convinced that if local farmers would plant potatoes their lives would be less in jeopardy. But neither persuasion nor demonstration won converts. So, as head of the army, he ordered every solder in the Duke's forces to plant potatoes, to care for, harvest, and eat them. The length of military service was sufficient in those days to give a soldier time to learn to grow potatoes and to develop a taste for them. "After the men returned to their farms and villages, potato crops appeared all over the country and the food of Europe gained greater security (Graubard 1943:5).

It must be remembered that the usual American attitude toward authority—that it must be legally achieved and judiciously exercised—is by no means universal. Power and authority for their own sake are respected in much of the world. A man may even be a tyrant, but his position is known, and his behavior is predictable. Awareness of this respect for power helps us understand why, in much of Latin America, the strong man, the *caudillo* who rules in arbitrary fashion, is often able to command a popular following. This is also true in other parts of the world. Ullah describes how, in a Punjab village in Pakistan,

> *The social goal of life of an individual and thereby of a family
> is to be effective and powerful enough to be of help to
> friends and awe the enemies. Success of a family and an
> individual in life is measured by the extent of one's influence
> over other people. A person who has no enemy, whom
> nobody fears and whom nobody obeys is a worthless person
> (Ullah 1958:171; my emphasis).*

Gibb and Bowen, speaking of the Arab provinces in the Otto-man Empire in the eighteenth century, make the same point:

*The conception of authority implied in the minds of the subjects themselves an assertion of power accompanied by a certain measure of harshness and violence. " 'Abd el-Ra'ûf Pasa (says the Christian chronicler Michael of Damascus) was mild, just, and peaceloving, and because of his exceeding justice the people of Dasmascus were emboldened against him."*

And in Egypt, quoting the Egyptian el-Cabartî:

*"If the peasants were administered by a compassionate multazim, they despised him and his agents, delayed payment of his taxes, called him feminine names, and hoped for the ending of his iltizâm and the appointment of some tyrant without fear of God or mercy for them" (Gibb and Bowen 1950:204–205).*

These extremes can hardly be justified in the modern world. But it can logically be argued that the use of authority to achieve directed culture-change should be a function of prevailing cultural expectations and administrative practices. This probably means that sometimes basic decisions must be made by persons in authority, de-cisions that in other societies might be handled through more dem-ocratic channels. I rather regret that in the city in which I live civic officials have not been able arbitrarily to fluoridate the water. Demo-cratically expressed public opinion sometimes is a poor way to establish truth.

In concluding this chapter, the fact should again be emphasized that people accept, and are able to accept, innovations because of a variety of interlocking and mutually reinforcing reasons. Frequently the reasons for acceptance are quite different from those that mo-tivate change agents. Dube speaks of how in India there is a ten-dency in cultures to reinterpret the proffered innovation in terms of the dominant themes and existing needs of the society. He points out that certain items in development programs, such as renovation of wells, paving of village lanes, and construction of soakage and compost pits, have been accepted in a number of villages where there is very little understanding of their significance for the health of the community.

*Their acceptance has been motivated by such diverse factors
as 'they look new and good,' and 'with them our village
will look like a town,' 'we must do what the government asks
us to do,' 'that is all that we can show the important
visitors from outside,' and 'other villages are doing it and so
we must also do it' (Dube 1956:30).*

Solien and Scrimshaw report that an experimental milk-distribu-
tion program in Guatemala was a relative success, but not for what
the program planners thought was the right reason. They found that
mothers feed their children well, not to *make* them healthy, but because they *are* healthy. It proved fruitless to try to persuade mothers
that milk should be given to children because it improves and main-
tains health. Nevertheless, milk was accepted for a variety of other
reasons, such as the fact that they liked the taste, they liked the
social aspects of the distribution, they wanted to cooperate with the
program personnel, and they feared that treatment in the much-
valued medical clinic would be denied to those who refused to take
milk (Solien and Scrimshaw 1957:100–101).

# 9

# bureaucracies
# and
# technicians

During the past generation most analytical studies of the processes of directed culture change have focused on the recipient group, on its social structure, its economy, its customs, and its values. In preceding chapters we have searched out the barriers to, or the inhibitors of, change that are rooted in individual psychology and in cultural and social forms, and we have looked for ways to neutralize or overcome these barriers. This is appropriate methodology as far as it goes, and in fact it represents an enormous step beyond the primitive conceptual frameworks that characterized early technical aid programs. In order to make clear the limitations of this methodological and conceptual position and to set the stage for the next advance, it is necessary to review approaches to technical assistance and international development.

## BUREAUCRATIC BARRIERS

### The primitive technical assistance model

Through private organizations, such as the Rockefeller Foundation and the Near East Foundation, technical assistance in the health, environmental sanitation, and community-development fields has had a history of more than 50 years. The Agency for International Development and its predecessor organizations (such as the Institute of Inter-American Affairs in Latin America) have carried on similar work in agriculture, education, and public health for more than 30 years. Comparable efforts on a multinational scale have been implemented since the end of World War II by the United Nations and its specialized agencies such as the Food and Agriculture Organization (FAO), the World Health Organization (WHO), and the Educational, Scientific, and Cultural Organization (UNESCO). The striking thing about *all* of the early work was its ethnocentric premise: *Techniques, programs, and solutions that have worked well in the most developed countries will work equally well in developing countries.* The first program planners and technicians assumed that technologies are absolutes, divorced from culture, equally suitable and efficacious in all sociocultural and economic settings. It was almost as if smallpox vaccination were used as a universal model for the transfer of *all* techniques, for here of course is a medical technique whose success has little or nothing to do with culture. Planners and technicians also assumed that the superiority of Western ways is obvious to anyone who observes them and that people in developing countries would be anxious, given the opportunity, to adopt them for themselves.

### The anthropological technical assistance model

These assumptions, and the premise from which they stem, were soon found to be an inadequate model on which to base worldwide technical assistance programs, but it is remarkable how long it was before a better model appeared. This "better model" is essentially anthropological; it postulates that the major problems in the diffusion of advanced technologies are rooted in the society and culture of the target group, and that programs and projects aimed at ameliorating undesirable conditions must be a function of, or adjusted to, the local cultures. The model assumes that people are anxious to improve their living conditions, and that they will modify their behavior *if* innovation is presented to them in such fashion that they perceive advantages and *if* the social cost in disruption of traditional and valued ways is not too great. Consequently, says the model, the recipient peoples must be studied in their social, cultural, and psychological dimensions so that (1) innovations will in fact represent functional improvements, and (2) they can be tailored and modified to conform to local cultural expectations.

## The donor-culture model

The anthropological model began to take shape about 1950; to the present it has represented the sophisticated approach to the problem. It did, in fact, permit great strides in the delivery of technical assistance. But we now realize that, just as the primitive direct transfer model was inadequate, so too does the "study the recipient culture" model reveal a major shortcoming. This shortcoming is its failure to consider the *donor culture*—the administrative organization, the bureaucracies, that plan and administer change programs and hire and supervise the technical specialists who are the direct lines of communication of new ways. The donor culture is, therefore, just as much a part of the process of directed change as is the recipient group. Unfortunately, methodologically and conceptually it has largely been taken for granted. Bureaucracies are a part of the daily life of almost all of us in complex societies and—mistakenly—we assume that through participation we understand them well, and that there is nothing particularly exotic about them. Consequently, because of this familiarity, only today are we beginning to realize that major barriers to change, and to efficient directed-change programs, lie in the structure and dynamics of innovating organizations and in the culture and psychology of the people who staff them. If refinements are to be made in our technical assistance methodology model and if developmental programs are to be more effective, it will be necessary to carry out institutional research comparable to that carried out on target cultures.

In this task we shall not be starting from scratch. Much excellent organizational research already has been carried out, especially by sociologists, on bureaucratic structure and professional cultures. But remarkably little effort seems to have been made to apply this knowledge, or to extend the objects of research to include the organizations that are engaged in international technical aid. More knowledge is needed, and this knowledge must be made operational; that is, it must be fed into the planning and operation of programs, just as we now feed knowledge of recipient cultures into planning and operations. And just as technicians are sensitized to cultural differences and their meanings for cross-cultural professional practice, so must they be sensitized to the bureaucratic world that provides their home base, and which in myriad ways, at conscious and subconscious levels, affects their perception of problem and mode of operation.

## Bureaucracies as cultures

In order to sensitize technicians and other bureaucrats to the deter-minitive role played by the organizational structure within which they work, we must begin with the recognition that a bureaucracy of substantial size is simply another kind of social grouping, with most of the same structural and dynamic features found in such "natural" communities as a tribe, a peasant village, or a city. Like these, it is composed of members of both sexes, with a wide age spread, organized according to functional tasks, in a hierarchy of authority, responsibility, and obligation. In other words, just as with a peasant village, a bureaucracy has a social structure that defines roles, relationships, and statuses of all members of the group vis-à-vis each other. New members continually are introduced into the system (through recruitment rather than birth); they are socialized and enculturated to accept the fundamental premises, values, and goals of the organization; they perform their professional assignments as long as they remain with the group; and through retirement, resignation, or separation (rather than death) they leave the organization, thereby making room for newer and younger members who are essential to organizational viability.

A bureaucracy resembles a natural community in other ways as well, including structural integration, institutional and individual behavior based on explicit and implicit premises, and personality and psychological variation among employees. As with all structurally integrated systems, no change comes about in isolation: New goals, new programs, new modes of operation imply rearrangement of role relations within the organization, bringing increased authority and status for some and lessened power and prestige for others. Bureaucrats, like all other human beings. jealously guard their traditional perquisites and positions, willingly surrendering vested interests only in exchange for something as good or better. Consequently, rearrangements in role relations, which favor some and threaten others, always meet with resistance. In bureaucracies this leads to organizational inflexibility, which makes it difficult to meet changing conditions and new needs. In other words, structural integration may be just as much a barrier to change in bureaucracies as in peasant villages. This problem is, of course, widely recognized in government and in private organizations; that it is so difficult to cope with is evidence of the monolithic quality of well-established cultural forms.

## Premises in bureaucracies

Less well recognized is the fact that cultural forms and individual behavior are—as pointed out in Chapter 1—functions of conscious and subconscious postulates or assumptions. In bureaucratic operations, these assumptions have an enormous influence in determining planning, administrative, and professional operations. Consequently, if we are fully to understand the dynamics of bureaucracies, and in the case at hand to see them in relation to planned-change programs, we must know what these assumptions, or premises, are. Only then do we appreciate the rationale for a particular form of behavior, a plan, a program, an administrative decision. Only when we recognize, and question, basic premises do we achieve the insight that permits fresh views of old problems, that leads to imaginative new approaches in all operations.

When we speak of the premises underlying American technical assistance organizations, we are simplifying, for there are in fact at least three levels of operative premises: general, national premises, largely traditional and middle-class in origin, which employees share with other members of their society; premises generic to the institution of bureaucracy; and premises specific to the several professions represented in bureaucracies. The *pastoral ideal* represents one widespread general American premise. Other commonly shared premises are the idea of progress and of *unlimited good* (in contrast to *limited good*), with the implication of room at the top for a great many, coupled with upward mobility. These, and many other premises of which they usually are dimly if at all conscious, color planners', administrators', and technical specialists' perceptions of problems and the strategies they devise to reach their goals.

Bureaucratic premises include the assumptions that an administrative organization must continually grow if it is to remain healthy, and that the importance of a position is determined by the number of people the incumbent supervises. A corollary of these premises is that it is preferable to fill an opening with a mediocre, or even a totally unsuitable, employee than to risk loss of the position. The consequences can be disastrous, particularly in international developmental work, where skill and sensitivity in human relations are especially important; one misfit can undo the work of a dozen well-qualified people.

Professional premises are many and varied. To illustrate, American medical personnel believe that human life is sacred and that no expense must be spared in an effort to save it. Agriculturalists seek the highest possible yield with the least possible manpower. Community developers often assume that, as Jackson puts it, "what counts is the way of tackling the job itself" (Jackson 1956:12). And educators believe that literacy is the key to modernization, the first priority in developmental programs. Whether these premises, and many others, are correct or incorrect is beside the point. What counts is that their mere existence will have an enormous influence in determining how technical specialists will view their assignments, shape their work plans, and provide scales against which to measure their success and that of their peers.

Against this background, it should be clear that in order to achieve maximum success in technical assistance programs, it is essential to realize that barriers to change are at least as prevalent within the innovating organization as within the target group. We need consciously to face up to the fact that bureaucracy places absolute limits on what we can do, that it burdens us with heavy costs, human and monetary, and that it must be studied just as thoroughly as a target group if results are to be optimized.

## THE TECHNICAL SPECIALIST

### Professional training and program orientation

With the background of bureaucratic structure, values, and premises in mind, let us turn to the question of the technical specialist, for he (or she) is usually the frontline person who has direct contact with national counterparts, and often with recipient people's themselves. How he sizes up his assignment, how he plans his work, how he relates to people in the country in which he is stationed—all of these factors are instrumental in determining how well his job is done. Many technical specialists have done superb work in foreign countries; others, regrettably, have been less successful; and all technical assistance missions record instances in which it would have been better for all concerned had a particular "specialist" remained at home. Sometimes specialists have failed because they were not, in fact, well trained; sometimes, because they lacked the human qualities or the adaptability to strange conditions that are essential for successful work.

In many other instances, paradoxically, specialists have been less successful than might have been expected precisely *because* of their enthusiasm, the excellence of their professional training, and the underlying presuppositions on which this training is based. Therefore, to understand the reasons for failure (and to be able to convert failure into success) we must know something about professional training. In the United States the first thing we note—and it is so obvious as to appear to have no significance—is that professional training is designed to equip the student to live and work in his own society. Or, putting the matter in another way, American society—like all other societies—is characterized by medical, engineering, agricultural, educational, and many other kinds of problems. These problems are all a function of American culture. Professional training, therefore, and quite rightly, is designed to teach people to meet the needs posed by these problems. There is no need to teach the average professional student to think first of identifying major problems and then working out solutions to them. The major problems are thought to be quite obvious; the need is to ameliorate, not to identify, them. The professional knows the kinds of questions his society will ask of him and the job demands it will make upon him. These are the tests he must pass successfully. Consequently, most professional training is designed in terms of *programs* rather than underlying *problems*, and the technician comes to judge himself by what he has to offer to the programs in which he participates. Professional training produces program-oriented specialists. Only rarely does it produce problem-oriented specialists.

However sound this approach is for professionals who will practice in the United States, it is seriously deficient in preparing people for work in other societies. Man is so much a product of his culture—he is so ethnocentric—that he assumes that the advanced programs and techniques that work in his society are equally fitted to less developed countries. The combination of fine technical training and an ethnocentric point of view leads to false and dangerous definitions of a good technical aid program and the role of the international technical specialist. The "good' 'program or the best technical assistance comes to be defined as the *duplication of American-style programs and projects*. The obvious corollary is that the best technical expert is the person who *most perfectly transplants an American-style program*.

Neither axiom nor corollary is true. In developing areas the answers to the major problems have not been worked out. The task of the technical specialist is not to reproduce a standard American product, but to know how to adapt the scientific knowledge and

operating techniques of his country to the economic, social, educational, and political reality of the country in which he works. The
successful technical expert is the one who has learned to be problem oriented and not program oriented.

### Program orientation in community development

The kinds of incongruities that may stem from rigid program orientation are seen in much community-development work. Since World
War II a great deal of successful village developmental work has
been achieved through programs variously labeled "fundamental
education," "community organization," and "community development." Not all work, however, has been entirely successful. The
bulk of community-development efforts and the underlying philosophy are an outgrowth of American and British experience. The
American pattern is one whereby people are stimulated to think
and talk about their problems, define their felt needs, come together for cooperative action, and through self-help programs, contribute to their material and spiritual well-being. Particularly in the
South, community projects based on this philosophy and concept
have been noteworthy.

When Americans participate in international community-development programs, their concept of the problem and their
program to attack the problem derive from the American model.
Consequently, the concept of American-style community development, and generally the attendant clichés as well, have been exported, much as health, agricultural, and educational programs are
exported. Many of the difficulties that have followed are due to the
fact that a concept is as much a function of a particular culture as is
a tool or a health program; it cannot automatically be grafted to a
foreign body. Let us examine, for a moment, some of the preconditions in the United States that were instrumental in defining
the philosophy and methodology of community development:

1. Communities have the power to tax themselves.
2. Administrative organizations with the legal powers to take
   action are under community control.
3. Populations are basically literate.
4. Leadership patterns are well developed.
5. Since the time of the frontier there has been a tradition of
   genuine cooperative work, and formal and informal social
   devices such as the town meeting and proliferating committees exist to implement this cooperation.

6. However depressed a particular small area, it is a part of a wealthy country, which in times of need will funnel help from other areas.
7. Technical services in health, agriculture, education, and the like are highly developed and available.

In short there is unlimited basic potential. The role of the community developer is therefore that of a catalyst, that of stimulating people to take stock, assess needs, decide upon action, determine priorities, and get to work. This makes sense in the American context. But let us look at the peasant villages in developing countries where this philosophy is being applied. Here are the corresponding characteristics commonly found:

1. The communities have essentially no power to levy taxes. They are at the mercy of national or state governments for all major and most minor developmental funds, including those for building schools and paying teachers. Any project that costs much money must be financed from outside the community.
2. Village government is truncated; only the most minor decisions can legally be made, and elected village leaders are reluctant to push beyond the modest limits of their authority.
3. Populations are not literate. People are often uncritical in their judgments, and rumors run rife.
4. Leadership patterns are poorly developed. Often the most competent people judiciously avoid entanglement in leadership squabbles.
5. There is little tradition for cooperation; there are fewer mechanisms for it; and people fear cooperation will enable their fellows to take advantage of them.
6. Peasant villages are part of economically depressed nations; they cannot count on the funneling of much help from other parts of the country.
7. Basic technical services are poorly developed, and sometimes lacking. Even if the community defines its needs, it cannot often get the outside support it wishes.

Hence, an American type of program based on the American philosophy often finds itself on rather barren ground. A community catalyst can go only so far. In fact, then, community-development programs in developing areas pay lip service to the slogans of

American community development—it becomes almost a religion—but "felt needs" usually turn out to be rather standard programs in environmental sanitation, medical services, agricultural extension, and education, which are recognized by national planners—correctly, I think—as the major needs of rural areas. It is not often that a "teahouse of the August moon" is built out of developmental funds because villagers insist it is their number-one need.

The point I am trying to make is not that community development is unsuccessful; it is good, and sometimes excellent, in spite of the natural handicaps under which it labors. The point is that a *conceptual* and *philosophical* cultural fit are essential in defining developmental programs, and that often highly competent technicians have not been prepared by their American training and experience to work out this fit.

When the American technician can discard his conceptual bias about what is good community development, he is in a position to put his skills to work. He will see that the first task is to consider the limiting factors in the total sociocultural-economic setting and assess such resources as are available. He will then realize that the major decisions as to what can be done in a community must be worked out on a nonvillage level, as a part of overall national social and economic planning. Perhaps the village worker will find he is less a catalyst and more a coordinator and explainer, one who helps villagers understand the role of substantive specialists in health and agriculture who formerly were not parts of traditional society. Perhaps his main task will be to prepare the ground so that the limited national services available can be most efficiently used on the local level. This will not be American-type community development, but it will be good community development because it is a function of the needs and possibilities, not of the United States, but of the local country.

### Direct transfer of American program models

The difficulties in which the program-oriented specialist may unwittingly find himself are illustrated by the following examples. In Iran in the early 1950s American public health technicians insisted, in the face of visible evidence to the contrary, that defecation in the open air would produce flies (dry atmosphere quickly dries fecal matter and flies do not breed). The program approach to this health problem, which logically had to exist, was the latrine, which, when installed in numbers, became a fly breeder, and villages previously free became infested. (communicated by Dr. Garegin Sarouk-hanian).

Technicians sometimes fail also to distinguish between technical excellence and cultural values, as the following vignette shows:

> *In my country [Iran] village public baths are pools of warm water. These pools transmit disease, and shower baths would be much more sanitary. Now without spending my time arguing about the goodness or badness of, or attacking or defending the ideas and behavior of my countrymen, you should know that they don't like to see themselves and others as naked as their innocent ancestors used to be in the jungle, and they will never retrogress, even for a few minutes in a public bath, to the way of living of prehistoric times. Because of this strong feeling men always wear something in public baths to prevent the lower part of their bodies from being seen by others. An American sanitary engineer built a public shower bath, in an Iranian village, but he didn't separate the stalls with partitions. I told him the design would not be acceptable because men would be ashamed to take off their clothes in the presence of others. He told me that they would have to accept it, because people are created alike and there is nothing to be ashamed of. Although perhaps he was right in his philosophy, the villagers did not accept his doctrine and the new bath house was little used. Moreover, they joked about the bath and the new ideology of human equality!*
> *(Communicated by Dr. Meydy Soraya.)*

### The temptations of overdesign

Another type of technician-centered problem, happily less common than one that seeks to duplicate an American-style program, is the specialist's desire to produce the perfect project. Technicians usually, and justifiably, are proud of their ability, of the high levels and standards of competence to which they have been trained; they are anxious to do the very best possible job, both for their own satisfaction and because it is gratifying to have one's peers recognize a superlative performance. Most professional work in the United States is limited by economic, political, and budgetary realities, and rarely if ever does a specialist have the opportunity to undertake the perfect project. The architect must settle for a less costly structure than the best he can design, and the highway engineer has limits placed on ideal routes and grade alignments by competing or conflicting demands for the same space. Medical doctors rarely have all the equipment and supporting staff and facilities needed for perfection in practice, and university teachers are hampered by class size, cost of supplemental materials, and other factors.

So it is not surprising that an overseas assignment has sometimes been looked upon as an opportunity to execute a project with a degree of skill and perfection impossible of achievement at home. At least in the early days of technical assistance, host country nationals often were dependent for design advice on the very people who were to carry out these designs. Thus, when decisions were made to build dams or to equip hospitals, they were at the mercy of foreign specialists who, for professional and ego-gratifying reasons, sometimes were desperately anxious to believe that the more elaborate the design, the better suited it was to the needs of the country. Overdesign, overly sophisticated projects, and ill-conceived programs sometimes resulted.

In 1957 in the Helmand Valley of Afghanistan I was shown irrigation works designed and built by American engineers and technicians to standards as high as could be devised on drafting tables. For all the technical excellence, however, serious problems were encountered in the local distribution of water to farmers' fields. American agricultural agents told me the degree of sophistication and responsibility needed on the part of the farmer in order to operate and maintain the superbly designed gates, valves, and ditches was higher than could reasonably be expected of American irrigation farmers. Yet many of the Helmand Valley farmers had only recently been resettled from a traditional nomadic life. The project designers had had an opportunity to design and build a system that, for a variety of reasons, they could not have built at home. The fact that their project was not an answer to local problems escaped them.

Even when there are highly qualified host-country planners and engineers who jointly share responsibilities with foreign consultants, the danger of overdesign is always present: They, too, have their models of technical perfection. This danger is illustrated by problems encountered in building the new *Ciudad Guayana* in Venezuela:

> *[The engineers] were anxious to proceed with the greatest possible speed* and to employ the highest engineering

standards, even where inappropriate. *For example, they cut unnecessarily wide swaths through woods for the construction of roads or utilities, and they made borrow pits wherever convenient, without regard for appearance, future land uses, or soil erosion* (von Moltke 1969:145; my emphasis).

It should be clear that neither inappropriate design nor excessively high standards are functions of nationality; they are the logical consequence of the *program* approach in education rather than the *problem-solving* approach, and almost anyone who has been trained in the former is apt to make the same error. This became clear to me when in 1951 I accompanied a well-trained, pleasant, young Latin American public health nurse as she made her assigned rounds in the poorest homes of the coffee plantations. Her instructions, which she had learned by rote from a translation of an American diet book, were to teach mothers to feed their children well, to emphasize oranges, eggs, meat, and milk. In house after house the advice was given, while unwashed, malnourished children played listlessly on dirt floors and the mothers numbly but respectfully promised they would follow instructions. These mothers considered themselves lucky to have enough beans and tortillas to dull the pangs of hunger; the recommended diet was, to them, nothing more than words. The *program*, which to the nurse was synonymous with good role performance, was to teach American standards of diet; the *problem* was to devise a way, within economic possibilities, to introduce some slight improvement in the traditional diet.

### Professional pride

Another source of difficulty to the technical expert lies in professional pride. Pride in the contributions one's profession makes to human welfare is a fine thing, but unalloyed enthusiasm for the magic in one's own field is equivalent to wearing blinders while navigating through shoals and uncharted seas. Good professionals come to feel that, although technological development is a complex thing involving many diverse efforts, somehow the contribution of their field is the key element, and that if they can only do what they feel is essential, most of the other problems will fall into line. The health specialist feels, "If only we can bring good health to these people, they can work efficiently, have more to eat, be stronger, and enjoy their leisure in a way not now possible." Quite true. The agriculturalist feels, "If only we can get these people to raise more and better food, they will enjoy better health and be able to work harder and enjoy their leisure." Also true. Educators feel, to a man, that the first thing to do in a newly developing country is to teach people to read and write; if you only make people literate, they argue, the remaining problems are not very difficult. And so on through the list of professions.

Professional compartmentalization—a function of American culture, with its great division of labor—tends to be exported along with professional knowledge. Sometimes no harm results. But, as should be clear by now, technological change is a multilinear, not a unilinear, process, and each activity builds upon, draws from, and contributes to many other activities. When professional pride is combined with professional jealousy, the joint planning and operations that produce the most successful programs are difficult to achieve. Professional indoctrination, it seems to me, often makes it difficult for a well-trained person to appreciate fully the contributions that are made by specialists in other fields and to realize that the success of these specialists will enhance the value of his own work.

### Ego-gratification needs of technical specialists

Technical specialists, as we have seen, are creatures of the premises and values underlying their national, bureaucratic, and professional cultures and subcultures, and their views of themselves as citizens and as professionals are deeply influenced by these forces. Every technical expert is also a unique psychological entity, a human being who, like all other human beings, must come to terms with himself and the world about him if he is to function successfully. The ways in which this adjustment is made are many, but an essential element is a person's ability to satisfy himself about his identity and his individual worth.

In professional fields the recognition we receive from peers, clients, and superiors is a major support to our self-image, to our view of ourselves as able and valued people. We like to believe that we are competent performers, and consciously and subconsciously we crave the recognition for our ability and performance that assures us that this is really the case. Frequently we are embarrassed to admit our need for recognition, since it suggests lack of personal modesty. Yet the need is there, and knowingly or unknowingly we manipulate a series of symbols in order to maximize the likelihood that recognition will be forthcoming.

In the business and industrial world, a recognition of merit comes through high salaries and bonuses; in politics, it comes through election to high office; and in sports, through winning. But among technical specialists—and university professors may be classed with this group—such symbols are of limited value. Our satisfaction comes from demonstrating "good role performance"— that is, from doing well what we are trained to do. The question then becomes, "What is good role performance, and how is it demonstrated?" The important thing to note is that usually it is not something that exists solely within an individual, without reference to others. Rather, it can be measured only in relation to other people. That is to say, *the response of those toward whom a professional directs his efforts is the critical element in assessing his performance.* A good respnse indicates a good performance, while a poor response suggests a poor performance.

As professionals, then, we are all very much concerned with how our client group—the people with whom we work—responds to our efforts. For a university professor, well-attended classes and eager students are highly gratifying symbols of good teaching, whereas poorly attended classes with apathetic students clearly indicate the opposite. For the American agricultural extension agent working abroad, local farmers who respond enthusiastically to his educational efforts are seen as evidence of superior performance, whereas lack of village interest suggests work of poor quality. Our clients— students or villagers—obviously have enormous power over us, since by their response they can grant or withhold the evidence of ability that is so important to us for personal psychological reasons, and so critical for our future careers.

When, as teachers or as overseas technical specialists, we encounter poor client response, we find it difficult to admit that the cause may be in ourselves. We see ourselves as dedicated, hard-working professionals whose primary interests are our students or our villagers. Technical specialists, particularly, may see themselves in a missionary role, with a sense of mission and urgency that leads them to live under difficult personal conditions for long periods of time far from home. When they encounter a less enthusiastic response to their efforts than they had expected, they feel angry and upset, emotions that when recognized, are often explained as due to disappointment that the client group members are jeopardizing their opportunities to make progress. Continued rejection all too often leads the technical specialist to project his feelings of inadequacy toward the people with whom he is working and to brand them as ungrateful, apathetic, or even stupid. Many years ago Embree noted that commanding officers in the U.S. military government on Saipan and Tinian were "obsessed" by the idea of gratitude:

> The matter of gratitude is a subject often present in administrator's minds and is to be found in many types of officer. They look at what they do from a personal point of view, they see how hard they work; they know they struggle to keep supplies moving, food distributed, houses built, health improved, and therefore they expect the people to appreciate their efforts. . . . the desire for gratitude is very great, and the lack of it hurts even the most understanding of administrators (Embree 1946:31).

In similar vein, Joseph has identified the rejection syndrome among American Indian Reservation physicians who find it difficult to elicit patient response because of cultural differences and misunderstandings occurring in the therapeutic interview.

> He [the physician] finds blank faces where he expects smiles, smiles where he expects seriousness, silence when he waits for an answer, and complaints where he expects approval and praise. . . . the physician, frustrated by the failure of his attempts at goodwill, and in his usual condition of physical exhaustion, sees himself as the victim of not being understood, as the target of secret attacks, and, rather often, as the "sucker" who is exploited by every patient. And out of this state of mind such statements as these are born: "Do not spoil them, don't be too friendly with them, because you will lose their respect" (Joseph 1942:3).

Even the most successful of overseas technical specialists will recognize that at times of frustration, or while unknowingly experiencing culture shock, they have been tempted to think such thoughts. More often than is realized, the specialist's success or failure is heavily influenced by his ego-gratification needs, and by his response to the vicissitudes of his daily work. Just as the recognition of culture shock—to be described in following paragraphs—is a major element in ameliorating its destructive effects, so is frank recognition of how egos are bound up in client responses a major element in promoting more effective professional work.

## Culture shock

Not only must a technical specialist make major professional adjustments in order to work successfully in a foreign country, he (and his family) must make major adjustments in their life styles and living conditions. Everyone, when first stationed in a foreign country, experiences *culture shock* to some degree. In the words of the anthropologist Kalervo Oberg, who first popularized the expression, culture shock is a malady, an occupational disease of people who have been suddenly transplanted abroad. Culture shock is a mental illness, and, as is true of much mental illness, the victim usually does not know he is afflicted. He finds that he is irritable, depressed, and probably annoyed by the lack of attention shown him by his local technical counterpart. Everything seems to go wrong, and the technician finds he is increasingly outspoken about the shortcomings of the country he expected to like. It rarely occurs to him that the problem lies within himself; it is obvious that the host country and its unpredictable inhabitants are to blame.

Oberg defines the symptoms of culture shock as excessive preoccupation with drinking-water, food, and dishes, fear of physical contact with servants, great concern over minor pains and skin eruptions, a hand-washing complex, fits of anger over delays and other minor frustrations, a fixed idea that "people" are cheating one, delay and outright refusal to learn the language of the country, an absentminded faraway stare (sometimes called the "tropical stare"), a feeling of helplessness and a desire for the company of people of one's own nationality, and "that terrible longing to be back home, to be able to have a good cup of coffee and a piece of apple pie, to walk into that corner drugstore, to visit one's relatives, and in general, to talk to people who really make sense" (Oberg 1954:2–3).

The malady of culture shock is caused in part by communication problems and in part by gnawing feelings of inadequacy, which grow stronger and stronger as the specialist realizes he is not going to reach all of those technical goals he had marked out.

*Culture shock is precipitated by the anxiety that results from losing all our familiar signs and symbols of social intercourse. These signs or cues include the thousand and one ways in which we orient ourselves to the situations of daily life: when to shake hands and what to say when we meet people, when and how to give tips, how to give orders*

*to servants, how to make purchases, when*
*to accept and when to refuse invitations, when to take*
*statements seriously and when not. Now these cues which*
*may be words, gestures, facial expressions, customs, or*
*norms are acquired by all of us in the course of growing up*
*and are as much a part of our culture as the language we*
*speak or the beliefs we accept. All of us depend for our*
*peace of mind and our efficiency on hundreds of these cues,*
*most of which we do not carry on the level of conscious*
*awareness (Oberg 1954:1–2).*

When a person enters a strange culture, these cues are re-moved; a series of basic props have been knocked out, and frustra-tion and anxiety follow. "When Americans or other foreigners in a strange land get together to grouse about the host country, and its people—you can be sure they are suffering from culture shock" (Oberg 1954:2).

Immunity to culture shock does not come from being broad-minded and full of good will. These are highly important character-istics of a successful specialist and they may aid in recovery, but they can no more prevent the illness than grim determination can prevent a cold. Individuals differ greatly in the degree to which culture shock affects them. A few people prove completely unable to make the necessary adjustment, and all technical aid missions of any size can point to skilled personnel who had to be repatriated because of in-ability to cope with local conditions. Other people get by with only a light touch of the affliction. Most of us go through a series of stages that represent a good, stiff attack of the illness, but we make a full recovery. Oberg has outlined these stages as follows.

During the first, or incubation stage, the victim may feel posi-tively euphoric. Probably he is staying in a good hotel or a staff guesthouse where food and sanitation approximate home conditions; the English language serves his pressing needs; the tourist sights are intriguing; the local people are courteous and helpful; and it is clear that a wonderful experience lies ahead. The new arrival notices col-leagues who have been on duty a few months who seem grouchy and depressed and ill adjusted, and he may feel condescending about these poor fellows who have not yet made the adjustment that he, the new arrival, has accomplished, all in a few days.

During this period he finds a house, maids, schools for the chil-dren, perhaps a chauffeur, and moves into the new home prepared to enjoy a scale of living he probably never knew at home. Then, wham! The Cook's tour is over, and the virus bites deep. There is maid trouble, school trouble, language trouble, house trouble, transportation trouble, shopping trouble—trouble everywhere. All the things about everyday living that were taken for granted at home now become insurmountable problems. The technician is now just another cog, as far as the bureaucracy to which he is assigned is concerned; he is no longer a novelty, and his national counterparts take him for granted. He is probably annoyed, too, because the gratitude he expects for his help is strangely lacking. The attitude is interpreted as indifference, or perhaps as an indication that the local people aren't friendly after all. At this stage the victim bands together with his fellow sufferers to exchange symptoms and to criticize the host country and all its citizens. The appraisal is derogatory, based on simple stereotypes that offer an easy rationalization for one's troubles. "These people can't plan," "They have no manners," "They ought to be taught how to get things done in a hurry," and so on, the complaining runs. At this period the cocktail circuit becomes a convenient crutch, an easy and uninhibiting atmosphere in which to get a load off the chest.

This second stage represents the crisis in the disease; if it is successfully weathered, the patient will be restored to health. Passing the crisis ushers the patient into the third, or recovery, stage. He now begins to understand some of the cues, which orient him, and perhaps enough of the language so that his isolation is not complete. Little by little the problems of living are worked out, and it becomes apparent that the situation, although difficult, is not absolutely hopeless, as it seemed only a short time earlier. A returning sense of humor is helpful at this point; when the patient can joke about his sad plight, he is well on the road to recovery. By now he almost imagines himself to be an authority on the country, and he can bolster his ego by talking in a knowing fashion before awed new arrivals. It helps, too, to realize that other people are experiencing the same depression and to be able to help them by holding out encouragement.

The fourth stage represents full or near-full recovery. By now, if ever, the technician will have made a relatively good adjustment to the situation in which he finds himself. He comes to accept the customs of the country for what they are. He doesn't necessarily wax enthusiastic about all of them, but he doesn't chafe. From time to time he experiences strain in his working relationships, but the basic anxiety of not being able to live is gone. Presently the technician realizes that he is getting a great kick out of the new experience and that there can be real exhilaration in the overseas experience. But however perceptive, no one realizes fully the nature of his illness until he returns to the United States on home leave, or again to live in that country. It is almost embarrassing to realize how many short-comings the good old U.S.A. seems to have and how frustrating and annoying so many experiences can be. Culture shock in reverse is much less serious than the original ailment, but it is surprising how many people can hardly wait through their home leaves to get back to the post that, only two short years before, seemed absolutely unbearable.

The difficulties that lead to culture shock are very real. A temperate-climate dweller has new health problems to face in the tropics, and food and water carry bacteria unknown at home to which an immunity must be developed. Business methods are different, and the corner shopkeeper may not agree that the customer is always right. Electric and water service may be cut at inconvenient times, and the telephone, even if it works perfectly, is a terrifying instrument when it carries a language foreign to the user. But these are usually minor difficulties. The environment remains the same, but the technician adapts himself to it; it is his changed attitude that has restored his health.

Culture shock is not limited to technical specialists on overseas assignments. Tourists all too frequently reveal the symptoms, and they respond like the stage-two technician: by griping and making unflattering comments about the country and its people. But there are two things that intensify culture shock for technical experts as against tourists and casual travelers. The tourist can always go home; usually he does before he has time to recover.

But the technician knows he is stuck for two long years, and the months ahead look like a life sentence. Not only this, but he has come to a foreign country with something of missionary zeal; the average technician is where he is because he feels he has something to contribute. Yet just at the time he is deepest in stage two of culture shock, he sees that he can't possibly accomplish in his two-year term the things he planned to do. The pace of the country seems slow; his counterparts seem uninterested; necessary materials are delayed; and budgets are held up.

The technician realizes, with horror, that he won't have much to show for his time. His self-esteem and his security are threatened, and the shock deepens. What will his professional colleagues back home—simultaneously his best friends and his most severe critics—think? In the final analysis our feelings of professional adequacy depend on how our colleagues evaluate us. We take pride in our ability but know that we must keep showing results if our reputation is not to falter. This means, unfortunately, that just at the time when we need maximum flexibility in coping with new conditions, security seems to lie in the course of maximum rigidity. The best technical expert, as pointed out, is the one who can appraise broad problems and decide on realistic courses of action. But program-oriented people need time and favorable conditions to learn problem-thinking, and this, while they are experiencing culture shock, is just what they don't have. In a strange and (apparently) hostile world, there is only one thing we can be absolutely sure of: We are first-class, A-1 professionals. But how do we prove it? Obviously, we demonstrate it, to our own satisfaction if not that of others, by trying to duplicate the stateside job that we have so often done before. The one thing that is not relative, in an apparently topsy-turvy world, we feel, is that there is a right way and a wrong way to do the job, and cost what it may, we're going to do it the right way. So the American sanitary engineer in Iran builds a bang-up American-type shower, and his colleague shows he knows how to cope with excreta-spawned flies—and the basic health problems remain untouched.

I recall a young American I encountered in a Latin American country, who appeared to be deep in stage two of shock. Slide rules stuck out of his pockets; pens and pencils were clipped to his lapels; his pockets bulged with tape measures; and he clutched drafting instruments in both hands. I knew immediately he was an architect. He was designing American-type small hospitals and health centers, using expensive imported materials and construction techniques suited to American building methods. Local architectural styles were pleasing to the eye and functional to the task; local building materials were inexpensive; the carpenters and masons knew how to build with them; and—with a problem-oriented outlook—better and cheaper buildings could have been constructed. But the architect had learned ARCHITECTURE, and he was making a good fight to preserve his sanity by convincing himself that he had learned his lessons well.

How long does culture shock last? It depends on the individual. Resilient people are over it in three months. Not infrequently it goes on for a year. Few people, when experiencing it for the first time, are well recovered in less than six months. A Brazilian health official once said, "It takes us a year to make a good consultant out of a North American technician," and this is a pretty good estimate of worldwide averages. A specialist begins to earn his keep after about a year of adjustment.

Does a case of culture shock immunize one against future attacks? Unfortunately, no. Future cases may be lighter and less frequent, but drastically different experiences can produce it time after time. In retrospect, I recognize at least three separate major attacks myself. On one occasion the primary cause, looking back, seems to have been loss of primary cues; on the other two occasions fear of failing on the job I had undertaken set off the attack. On the most recent occasion I knew all about culture shock, but it was not until some months afterward, when I looked back on the experience, that I realized what accounted for my behavior.

It is curious that culture shock has not been formally recognized until very recent years. Although there is no immunization that will prevent it, simply knowing that it exists, that an overseas assignment is apt to bring it on, and that it is not permanent, in itself reduces the severity of the attack. The technician who has been told to expect culture shock weathers the ailment much more easily than the one who doesn't know he is ill.

# 10

## the anthropologist
## at work:
## the conceptual
## context

Technological development programs run more smoothly and are more successful when the sociocultural patterns and the premises, values, and motivations of the members of both the innovating bureaucracy and the recipient peoples are known, and when the social dynamics of the project setting are understood and utilized in planning and operations. If this axiom is accepted by program planners who have the resources and the desire to gather such information, how do they go about getting it? There are some program planners and technical specialists who seem to know almost instinctively what they can and cannot do in a given situation, and their sensitivity is heightened through field experience. But such people are the exceptions; most program planners and field technicians do not have an innate feel for the social implications of guided change, nor are they fully aware of the possible consequences of their work. The ethnocentric blinders with which our culture provides us have to come off little by little. The sociocultural dimension in planned change is a little like grammar in language; it is there all the time, and basic, but it is not obvious until one's attention is called to it. Even when it is recognized, it is mastered only after hard study.

Anthropologists and other behavioral scientists have done a

great deal of research on the recipient peoples in programs of directed culture-change, and representative examples of that work are reported throughout this book. In making these analyses we have also considered in some detail the interaction setting—the dynamics of the project itself—where innovators and recipient peoples come together. Consequently, we have a pretty good idea how to go about gathering the necessary information and how to pass this information along to the ultimate consumers: the planners, administrators, and technical specialists. Regrettably, as pointed out in the preceding chapter, we have had almost no experience in making comparable studies of innovating bureaucracies, and even less success in convincing planners and administrators that these analyses can be highly constructive aids to their work. There is not, to my knowledge, a single analysis of an American technical aid mission in a foreign country, nor have intensive studies been made of any part of the Washington operations of the Agency for International Development.

One does not have to look far to realize why so little research on innovating organizations has been carried out. In the first place, even among many behavioral scientists, there is little appreciation of the practical importance of giving equal weight to the study of technical aid bureaucracies. In the second place, it is only human to assume that problems encountered in developmental projects are "out there" in the recipient cultures. The professional worker in an action program has little difficulty in seeing that the culture of the group with which he is working will determine how people react to him, in many ways. He can see that patterns of authority and concepts of role in the client group will affect the ways in which he must direct messages or ask for cooperation. Tolerance and objectivity in studying others is not hard to achieve.

But it is not so easy for a technician or an administrator to accept the fact that an understanding of *his* attitudes, values, and motivations is just as important in successfully bringing about change. It is painful to realize that one implicitly accepts assumptions that have little validity beyond tradition and that one's professional outlook has been uncritically acquired. It is not easy for either the technician or administrator to admit that the way he views his assignment and how he works are conditioned by such things as how he perceives his role in the bureaucracy to which he belongs, how he reacts to his supervisor, and how he deals with his colleagues and those who work under his direction.

In other words, bureaucrats are not anxious to believe that they can profitably be studied. Nor are they, in most instances, anxious to be observed and questioned as they go about their daily chores. Even when they are interviewed with great tact, and even when they have been well briefed as to why a study is being carried out, friction and hard feelings may develop. For some of them the mere fact that questions are asked implies the suspicion of an unsatisfactory performance; this is especially true of those who know that anthropologists often have been critical of technical workers. For most employees in a bureaucracy the line between performance evaluation and disinterested research seems rather fine.

For these reasons this and the following chapter are based largely on anthropological analyses made of recipient cultures as a part of directed-change programs. These are the sources that we must tap to extract general principles about change and to draw up guidelines about how social scientists can be most effective in such programs. In the future, I hope, we will have the experience necessary to describe how the anthropologist works among his social and professional peers, with program planners and administrators, in the total context of the bureaucracies of which they are a part.

## PROTO–APPLIED ANTHROPOLOGY

Working in a developmental program, the behavioral scientist formally utilizes the theoretical concepts, the research methodologies, and the factual data from the fields of anthropology, sociology, and psychology to facilitate change in goal-directed projects. The expression "formally utilizes" is the key to what is unique in the behavioral science approach, because throughout history men of great insight have utilized cultural, social, and psychological knowledge to achieve change in the behavior of groups of people. They were not making *formal* use of specific knowledge, but they knew what to do to accomplish their aims. Some of these examples make revealing reading for those of us who work in international development programs.

In the Spanish conquest of America, Christianity was successfully implanted in most areas because the church was familiar with the strategy of directed culture-change. Friars and priests learned the Indian languages in order to communicate with their charges and to study the pagan forms they wished to extirpate. They built churches on the sites of former Indian temples, and they encouraged the identification of pagan gods with the Virgin Mary and the saints. In these activities the church was merely following the lessons it had learned in earlier centuries in Europe, when, during the spread of primitive Christianity throughout the Mediterranean, pagan deities often were transformed into new guise through the process of syncretism.

The Venerable Bede records one of the earliest examples of a sociocultural tactic for directed culture-change. In *The Ecclesiastical History of the English People* he tells of the reconversion of England at the end of the sixth century by missionaries sent from Ireland, and he reproduces a portion of a letter from Pope Gregory the Great to the abbot Mellitus in 597 or 598 that might serve as a modern text on how to utilize cultural and psychological knowledge to achieve one's ends.

> *When then God shall bring you unto our reverend brother Augustine bishop tell him, what I have of long time devised with myself of the cause of the English men. That is to wit that not the temples of the Idols, but the Idols which be in them be broken, that holy water be made and sprinkled about the same temples, altars builded, relics placed. For if the said churches be well made, it is needful that they be altered from the worshipping of devils into the service of God: that whilst the people doth not see their temples spoiled, they may (forsaking their error) be moved the more oft to haunt their wonted place to the honour and service of God. And for that they are wont to kill oxen in sacrifice to the devils, they shall use the same slaughter now, but changed to a better purpose. It may therefore be permitted them, that in the dedication days or other solemn days of martyrs, they maketh them bowers about their churches, and feasting together after a good religious sort, kill their oxen now to the refreshing of themselves, to the praise of God, and increase of charity, which before they were wont to offer up in sacrifice to the devils: that whilst some outward comforts are reserved unto them, they may thereby be brought the rather to the inward comforts of grace in God. For it is doubtless impossible from men being so rooted in evil customs, to cut off all their abuses upon the sudden. He that laboureth to climb up unto a high place, he goeth upward by steps and passes, not by leaps (Hereford 1935:56–57).*

Pope Gregory showed acumen in protecting himself against possible criticism within the church by following the rule, as valuable today as then, of justifying proposed change by citing religious precedent or authority:

> *So unto the children of Israel being in Egypt our Lord was well known. But yet he suffered them to do sacrifice unto him*

*still in offering up beasts unto him, which otherwise they
would have offered up unto the devils, as they were wont to do
in the land of Egypt, that altering their intent, they should
leave some, and also keep some of their old sacrifices: that is,
that the beasts which they offered before, they should now
offer still. But yet in offering them unto the true God, and not
unto the devils, they should not be the same sacrifices in
all points as they were before (Hereford 1935:57).*

More than a thousand years later a Spanish premier showed a
comparable knowledge of culture and psychology in achieving a
change that stymied the forces of law. On various occasions in Spanish history the government, as an aid to crime detection, tried to
prohibit men and women from covering their faces with cloaks, hats
with turned-down brims, shawls, or scarves. In 1766 King Charles III,
at the instigation of his unpopular Sicilian Premier Squillaci, dic-
tated a Royal Order prohibiting soldiers and government employees
from wearing the long cape and broad-brimmed hat, an order subsequently extended to the general public. The resulting furor is known
in history as the Mutiny of Esquilache, and it resulted in banishment
for the hated foreigner. His Spanish successor, the Count of Arandas,
while in sympathy with the restrictions on dress, rescinded the order.
He accomplished its purpose easily and painlessly by making the
long cape and broad-brimmed hat the official uniform of the public
executioner (Altamira 1949:443–444).

In 1857 the Anglican missionary William Duncan began work
among the Tsimshian Indians on the cost of northern British Columbia. He quickly realized he must understand, not only the language,
but also the culture and the system of interpersonal relations that
structured this society. His approach is particularly noteworthy because of its recognition of the problems of perception across cultural boundaries. He realized that there were aspects of Christian
metaphysics that, if adhered to literally in preaching and services,
might be interpreted by the Indians in a very different way from that
intended. Since some of the religious practices of the Tsimshian involved cannibalistic rites, Duncan refused to introduce the sacrament
of the Lord's Supper. He feared the distinction between symbol and
substance would not be easy to impart. Furthermore, since he was
trying to eliminate the drinking of alcohol among his followers, he
felt that giving them wine in a church service would confuse them
and prejudice his work. Again, because of the strong Tsimshian emphasis on status and social classes and the outward symbols of posi-

tion, all of which he disapproved, he refused to wear the vestments of an Anglican priest for fear he would encourage rather than discourage social differences. He likewise eschewed religious paintings and carvings, believing that the act or symbol would assume a talismanic virtue in the eyes of his followers.

Duncan's unorthodox ideas alarmed his superiors, and in 1877 he was replaced by a regular clergyman with less cultural insight. Duncan's wisdom was demonstrated almost immediately. He had de-emphasized the concept of the Holy Ghost and the acceptance of direct revelation because of the importance among the Tsimshian of bodily possession by mythological spirits, which caused the possessed individual to act in a rapturous and often inhuman fashion. Shortly after his successor began preaching a more orthodox Christianity, several converts became ecstatic, declaring they had witnessed miracles, had seen Christ on the cross, and had conversed with the Holy Spirit. The contagion spread to other settlements, and the Indians sang and danced. At the height of their ecstasy some offered to give the power of God, which they held in their cupped hands, to anyone who wished it, a direct transfer of the native pattern of inducing spirit possession in a cult initiate (Barnett 1942).

## CONTEMPORARY APPLIED ANTHROPOLOGY

In contrast to these examples of proto–applied behavioral science, the modern anthropologist or sociologist working in a goal-directed program consciously utilizes the substantive data, the theoretical concepts, and the research techniques he controls. In a specific project the anthropologist—to draw from the author's field—tries to analyze the basic characteristics of the society and culture to which he has been sent, to understand the premises, motivations, and values of the members of the bureaucracy that is attempting the change, and to learn the patterns of interaction of the two systems. He brings a body of theory, which provides the framework for his analysis. Using this theory, as well as the factual information he has or acquires, he attempts to predict what the total range of change will be if a specific program is successful, or, conversely, he attempts to determine what minimum changes in an entire culture must occur before a specific project can succeed. He looks for the social and cultural barriers that inhibit change, suggests ways to neutralize them, and decides what motivations and other stimulants to change can most successfully be used in direct action, in aiding people to decide whether they wish to accept or reject proposed innovations. He hopes to help the members of the innovating group understand better the implications of its activities and the ways in which its organization and values bear upon the task at hand.

But how, more precisely, does the anthropologist go about his task? How does he progress from the general to the specific? How does he translate broad cultural and social knowledge into the concrete terms that can be utilized in a program of international public health or community development? The first thing to note is that there are no precise rules for the application of anthropological knowledge in such situations. There are no do-it-yourself guides that tell what particular bits of theory or fact are significant for specific problems. Intuition, the ability to sense problems and mode of attack, is essential to the behavioral scientist who is to work successfully in an action setting. In all fieldwork there is an artistic element, which must be present if the results are to be significant. The effective applied anthropologist must be well grounded in social and cultural theory; he must command a substantial corpus of factual knowledge; and he must be sensitive to the widest possible range of stimulants in his fieldwork.

### Three anthropological contributions

With this caution in mind, it may be suggested that the anthropologist contributes three important things to any program: (1) a point of view, or a philosophy; (2) factual knowledge; (3) research techniques appropriate to the task.

The point of view is something like this: No culture is all good or all bad. Basically, all cultures are reasonably good; otherwise they would not have survived. The anthropologist is conditioned by training, and perhaps temperament, to look for the good in a society, goodness being defined in terms of the society's ability to satisfy the needs and aspirations of all its members without jeopardizing those of other societies. The anthropologist is not opposed to change, but he does not necessarily approve of change for its own sake. He believes that real progress is made when it springs from and builds on the good things already existing in a culture, rather than when it is defined in terms of an approximation to the American way of life.

This point was illustrated in a seminar in the School of Public Health at the University of California, where we spent several sessions on the problem of diet in relation to culture. Students from foreign countries, studying nutritional problems, outlined for us the typical diets of their countries. Compared with National Research

Council standards, the pictures were not encouraging. There were grave shortages of such foods as milk, citrus fruits, meats, and vegetables. In view of the economic limitations generally prevailing, most students felt that adequate diet in their countries was something that could not be attained for many years. Then, with the aid of Dr. Ruth Huenemann, Professor of Public Health Nutrition, and without direct reference to National Research Council standards, we looked for ways to improve the diets, using only those foods that grew or could easily be produced in each country. We quickly discovered that, although there were serious weaknesses in all diets, none was really bad. All could be greatly improved within the economic limitations of each country by building on what existed, rather than by trying to approximate the standards that had been developed for Americans in the United States. Viewed in this light, a seemingly hopeless problem became quite susceptible of solution.

The factual knowledge the anthropologist brings to an action program is both theoretical and substantive. It consists of a body of general theory about society and culture, and particularly about processes of change—such as those described in earlier chapters— that have wide cross-cultural validity, and it consists of concrete data about the patterns of culture of the area in which the work is to be carried out. For example, the anthropologist who joins a public health or community-development program in Latin America is first of all a general anthropologist. In addition, he presumably has read a long series of monographs and articles on Latin American culture, knows something about the history and geography of the area, and in general understands the basic characteristics of the societies and cultures of this area that set it off from Africa or Asia. To the extent that good basic anthropological research has been carried out previously in a project area, the anthropologist is able to proceed to specifics with just that much more dispatch.

The research techniques the anthropologist brings to bear on field problems are the standard social science methods of interviews, schedules, censuses, and questionnaires. Of these the most important is what the sociologist calls the "open ended" interview and what the psychologist calls the "depth" interview. It is an intensive rather than an extensive method. It is based on the assumption that detailed data from ten informants give more meaningful answers than superficial data from one hundred. Further, it is based on the assumption that the range of factors affecting a problem can never be fully defined in advance. The way to find out what is significant is to find informants who are willing to talk, to guide them gently, but for the most part to give them free rein. Only after the range of a problem has been explored in this fashion is it profitable to attempt to quantify data by means of a questionnaire. Is the problem the lack of public response to a new health clinic in a depressed area? The anthropologist's approach is to knock on doors or otherwise seek introductions to a small number of homes. Long hours of general conversation, guided but not forced in the direction of the clinic, eventually will afford information on health beliefs and attitudes, authority patterns within the family, working schedules of the parents, economic problems, and a host of other factors that may be significant. Only after the potential breadth of the problem has been thus defined is it worthwhile to frame schedules and quantify responses.

### Initial steps

In approaching a specific assignment, the anthropologist tries to do two things.

1. *To determine the relation of the institution or elements involved to the total culture pattern.* Remembering that a culture is an integrated, functional unit, he tries to find out the nature of the phenomena he is studying, how they interlock with the remainder of the culture, and what their role and function are. In a health program, for example, he first asks what the concepts of illness and health are, and what role or roles are played by illness. In a housing program he asks what the nature and function of a house are, what it means to its dwellers, how it relates to family structure, economic activities, hospitality patterns, and the like.

2. *To determine the patterns of interpersonal relations among all people who participate in a program.* This means a knowledge not only of the social structure of the group, but also of the new relations that develop between the group and the organization that is trying to bring innovations from the outside.

The anthropologist begins by drawing upon his general knowledge. In a general way he is prepared to say what the range of functions of housing, of the family, of illness, of agriculture, of literacy, may be in a sociocultural setting of a given type. But this general analysis will take him only so far. It must be followed up by fieldwork to determine the specific and unique characteristics of a situation. Only then can recommendations safely be made.

This process can be illustrated by a hypothetical project designed to improve substandard housing. The anthropologist begins by asking very general and perhaps obvious questions: What is the nature of housing in this area? What is the function of a house? What are the purposes of housing thought to be by the people? Some of the general answers that come to mind are: A house is for shelter; it is a place to store food and feed a family; it provides bathing and toilet facilities; it affords privacy; it is a place for family interaction in work and recreation; perhaps it gives aesthetic satisfaction; perhaps it is a device to achieve prestige; perhaps religious functions are served there. These answers, and many others, can be worked out in a moment. But if we really want to understand what a house is, and what its social, economic, and other functions are, we must study concrete examples and note the ways in which each room and facility is used, the time devoted to each, the motor patterns associated with the use of equipment and furnishings, the relation of work to recreation areas, and a whole series of similar things.

### A Mexican case study

The manner in which general appraisal of a problem in housing can be worked out in the field is illustrated by an analysis directed by Isabel Kelly. In Mexico the Ministry of Health and Public Assistance wished to improve housing in a number of *ejido* communities in the La Laguna region of the northcentral states of Coahuila and Durango. *Ejidos* are rural communities holding communal lands distributed following the Mexican Revolution that began in 1910. Sometimes lands are farmed communally, but more often plots are distributed on an individual basis for each farmer to work as he wishes.

Kelly and her Mexican associates carried out the major work over a period of seven months in the *ejido* of El Cuije, about 15 miles from Torreón. This settlement was found to consist of 57 *ejidatario* families (with land rights), totaling 356 persons, and about the same number of families without rights. The village was of irregular plan. Houses were adobe, with earth floors; roofs were made of mud and reeds over beams; windows were small, with board shutters and no glass. Sanitary facilities were poor: Only half of the houses had water taps in the patios, and the remaining families used public fountains. Latrines were completely lacking. Garbage and rubbish were simply thrown into the street. One house had a cold-water shower, but otherwise men bathed in irrigation ditches and women in tubs in their houses. Small domestic animals walked in and out of houses at will.

Since Kelly had lived in Mexico for many years and had studied

Mexican culture in a number of local situations, and since her colleagues were Mexican, the group knew a great deal about basic Mexican culture. They were also imbued with the philosophy of building upon what already exists. But no anthropological research had been carried out in this particular part of Mexico, and no research directed toward housing problems as such had been recorded in any part of the country. Hence, a significant amount of research was necessary. During the first ten days the anthropologists visited homes, made friends with a number of people, and in a series of open-ended interviews began to block out the parameters of the problem. They learned a great deal about the nature and functions of houses and the social uses to which they are put. These observations served as guide in working out a detailed questionnaire, which supplied the statistical frame for analysis and recommendations. But it quickly became apparent that much more than a questionnaire on housing was needed, and ultimately the studies embraced data on farming techniques and attitudes, care of domestic animals, family budgets, nutritional problems, clothing, division of work within the family, traditional medical beliefs and practices, political and social organizations within the community, relations of the *ejidatarios* with the national government, and a whole series of similar things, which at first glance have little or nothing to do with housing. A solution to housing problems, it turned out, required knowledge of a great chunk of culture.

The analysis of the 57 houses studied revealed significant data. In 45 the kitchen served also as dining room, in 13 as living room, in 10 as bedroom, and in 6 for crop storage. In those houses with living rooms the space was often used for sleeping, sometimes for eating, and sometimes for crop storage. Obviously, multiuse of rooms was a basic pattern to be considered in planning improved quarters. In furnishings, first priority was given to at least one bed; this was found in all 57 houses. Second priority was given to a wardrobe, found in 85 percent of the houses, and third priority to a dining-room sideboard, found in 75 percent. Sixty-six percent of the houses had radios, and 50 percent had sewing machines. Half the kitchens had kerosene stoves, and all had elevated adobe hearths, usually with hood and chimney. Other kitchen furnishings included the metate (grinding stone), water jars, a few shelves, pottery on the walls, knives and spoons, and the like. Kitchens usually presented a considerable state of disarray. Clothes were washed in irrigation ditches or in wooden tubs in the patios. All families had domestic animals, which not only had practical utility, but served also as savings (that is, they could be sold in time of emergency) and as items of prestige, especially in the case of horses. Maize was stored in temporary cribs erected in patios, and agricultural tools were thrown in any vacant space on the property.

In addition to the analysis of house composition and uses, all 57 families were asked what they thought the ideal house should be, always given the economic limitations present. This revealed how important it is to work with, and not to plan for, people. For example, the anthropologists thought that running water in kitchens would be indicated because of the convenience afforded, but wives were almost unanimous in preferring a tap in the patio. Faucets leak —this is a fact of life—and a leaky faucet in the kitchen could be a real problem. Again, it seemed to the anthropologists that permanent maize storage facilities would be desirable, but it turned out the people preferred the demountable type, since this afforded more yard space when supplies were depleted.

With all the data in, general recommendations were made to the Ministry:

1. Houses should be oriented to face the southeast. They thus backed against the cold and wind of winter and the hot sun of summer afternoons.
2. Lot sizes should be 20 by 30 meters. This was recognized as smaller than desirable, but the people preferred a bit of crowding to encroaching on valuble agricultural land.
3. Traditional building materials should be used, because of economy and local knowledge of construction, but, with cement floors and larger windows on house fronts, better sanitary conditions would exist.
4. The minimum unit should consist of a kitchen-dining room, a living-bedroom, and an outdoor covered porch. This *corredor* would serve as a third room: It is ideal for much work, and in good weather some family members could sleep on it.
5. With respect to sanitary facilities, it was recommended that a single tap be installed in patios, for the reason indicated, and that water be distributed within the property with a hose. For washing and bathing, it first seemed desirable to combine a shower with a laundry room, since both required water and a drain, and times of use did not overlap. But it turned out that these facilities were in no way connected in the minds of the people. Further, women like to wash outdoors, under a shade, and not inside a small damp room. So in the

final recommendation provision was made for a small shower room with a water tank on the roof, where water would be slightly warmed by the sun. Clothes were to be washed in a simple concrete tub under a shade. A pit latrine was recommended, with a simple septic tank for water drainage.

6. The kitchen plan was modified to provide more storage and shelf space, and counters of different heights for different uses. For example, grinding with the metate requires, for comfort, a support at a height midway between the floor and the raised adobe hearth.

This basic design was flexible, in that additional rooms could be added as needed and as finances permitted; it made use of traditional materials and building skills; it met basic social needs; and it did no violence to any known values. Improved housing, designed on the basis of this kind of information, is much more apt to be successful than if the problem is thought of simply as an architectural and economic one (Kelly 1953).

## RESEARCH ON HEALTH PROBLEMS

### Social roles of illness

Several examples from the field of health will further illustrate how the anthropologist goes about his task and how he analyzes the cultural, social, and interpersonal factors involved in specific situations. The most comprehensive of the studies to be considered was made by Margaret Clark in a Mexican American enclave in the city of San Jose, California. This study was planned as an experiment to see how social science knowledge might be used to make public health education programs among minority groups in the United States more effective, and in general to point up the existing problems and conflicts that make medical and health needs among such people more difficult to solve than they are among native-born American whites. A committee, consisting of a professor of public health, the director of public health education of the state, the director of the county public health department in which the work was carried out, and three anthropologists, was set up to direct the work. The committee met regularly during the eighteen months of the study to hear progress reports, to make recommendations for further work, and to ensure the closest possible integration of health and social science interests.

It was recognized that health, illness, and the mechanisms to restore one to health are intimately related to the entire culture of a group. Consequently, Clark first devoted her efforts to making what is known in anthropology as a "community study," in which the basic patterns of life are outlined: social structure, economic activities, religious forms, relations of the group to local government agencies, basic attitudes and values, and the like. Only against this background, it was believed, could health problems be adequately analyzed. Next Clark made a thorough study of the ideas about sickness and health. What is health? How is it defined? How is it maintained? Why do people fall ill? What is done to restore the sick to health? Who are curers and what are their training and techniques? What part does the family take in making decisions about medical treatment? In general, the answers to these questions fell in line with the broad patterns of folk medical belief and practice in Latin America, which have been well described in recent years. But the Sal Si Puedes group lives, and has lived for a long time, in an American setting, exposed to urban American health services and sanitary laws. It is gradually acculturating to an American way of life, although the progress is uneven and halting. This meant that knowledge about Latin American ideas of health and disease was useful in guiding research, but the significant answers came from the local fieldwork.

One of the most important areas of data had to do with the manner in which illness was found to help stabilize social relations within the community through publicizing and punishing social offenses, providing a socially approved escape from censure for unsanctioned behavior, and dramatizing the acculturative situation. In one instance a young wife expecting her first child scolded her husband for returning home drunk. He beat her and put her out of the house in the rain. Her parents took her to the local curer (*curandera*), who treated her so the unborn child would not suffer "fright illness" from the experience. Under normal circumstances this domestic spat would not have attracted attention. The husband had availed himself of a male prerogative—an evening out with the boys—and his wife had humiliated him by her scolding. Had she not been pregnant his beating would have passed unnoticed. But because he endangered the life of his unborn child, the wife gained the support and sympathy of the community. Finally he saw the error of his action, apologized, and returned to his wife.

In another instance a woman with six children was faced with a prolonged visit by her husband's unemployed brother, his wife, and their five children. With fifteen people living on the wages of a single laborer, the debt at the grocery store grew larger and larger and finally credit was cut off. With Mexican patterns of hospitality and mutual obligations between family members deeply engrained in her, she recognized and accepted her obligation to her husband's relatives. Yet she feared her children would not have enough to eat, and the crowded conditions of the house had become almost unbearable. She began to suffer rapid pulse, shortness of breath, and sweating, symptoms that were defined as "fright" by the *curandera*. Now relieved of the obligation of caring for her brother-in-law's family by a socially sanctioned condition—illness— there was no alternative but that the visitors should move on. Had she remained in good health, she would have been thought to be selfish and inhospitable if she had complained about her husband's relatives.

Illness may also be a protest against the acculturative situation. An elderly woman admitted to a hospital against her wishes was required to take daily showers after the acute phase of her illness was past. She objected, since her custom was less frequent tub baths, but her objections were ignored. While returning from her shower shortly thereafter she had an attack of "bad air," a folk-defined illness, which, in her opinion and that of the family, was the result of the dangerous practice of daily showers. The illness gained her the sympathy of her family, who caused her to be released prematurely to be taken home for what they felt was the proper care (M. Clark 1959:198–202).

Upon completion of her study it was possible for Clark to make much more intelligible to county health personnel the reasons for the health behavior of Mexican Americans. Furthermore, she was now in a position to make a series of specific recommendations, which, to the extent they could be carried out, should alleviate some of the problems plaguing workers in this intercultural health situation. These recommendations had to do with communication problems, economic problems, the conflict between modern medical practice and folk beliefs, and problems relating to the definition of disease, to modesty, to hospitalization, and to differentially perceived medical roles (M. Clark 1959:218–219).

### Supportive roles in a rural Greek hospital

The importance of understanding the complex interpersonal relations that exist in any setting is illustrated by Friedl in a brief article about a hospital in rural Greece. Traditionally, she says, hospitalization has been viewed as desertion of the sick person by his family; in spite of this attitude, in recent years small private hospitals of from 10 to 25 beds have sprung up in rural Boeotia, and hospitalization is increasingly sought by villagers for childbirth and serious illness. The physicians who own the hospitals have been trained in Athens, and many have studied outside the country. Nurses, however, are local girls who have been trained by the doctors.

Friedl gives us a word picture of a sickroom: There is an iron bedstead against each of the three walls away from the door, covered with linens brought by the woman who occupies the bed. In the center of the floor of the 12-foot-square room, five relatives of one patient are seated around a blanket on which are the remains of a picnic lunch. At the head of the bed of a young woman patient her husband is heating macaroni over a burner on a small table; when it is warm, he feeds it to his wife. The husband and daughter of the third patient stand by her bedside, watching the goings-on in the room. Two of the patients wear their own nightclothes, while the third is fully dressed, lying on her bed.

The informality of such hospital treatment is a far cry from the modern hospitals of Athens and yet, Friedl reminds us, the technical and social patterns portrayed are almost ideally suited to the problem of introducing improved medical care to rural areas.

> These hospital scenes and the patterns of hospital care they
> represent suggest a remarkable similarity between the treatment
> of illness at home and its treatment in a hospital. The same
> values and attitudes of Greek culture are demonstrated
> in both situations, and their reinforcement in a time of stress
> is not impeded by hospitalization (Friedl 1958:25).

The Greeks feel human companionship is an absolute good; solitude is unpleasant and to be avoided even when people are well. Therefore, when someone is in the vulnerable state of illness, it is particularly important that he be accompanied at all times by relatives. Since the family has such important social and psychological functions, "the presence day and night of family members in hospitals fulfills the latent function of emotional support. Such support is essential for Greek patients, because they feel useless and unwanted whenever an illness prevents them from fulfilling their customary roles in the household" (Friedl 1958:26). The care by family members, the home cooking, and the crowded room—all anathema to a Western-trained nurse—all help to lessen the potential sense of isolation and strangeness of hospital surroundings. Friedl points out that these improvised hospital practices conform in many ways with recent social science recommendations for more flexibility in traditional American methods of institutionalizing hospital care. "What careful analysis has pointed out as a desirable method for conscious, planned, and gradual change, has been evolved willy-nilly by these [Greek] doctors under the diverse pressures of their complex culture" (Friedl 1958:27).

### Social relationships in an innovating organization

The final example shows how the network of social relations in the innovating organization affects a program, and how administrative changes may be reflected in the impact of a program. In a fine large health center in Santiago, Chile, with a wide spectrum of curative and preventive services, the director decided that a well-planned program of health education was the next step to bringing better health to his area, the population of which was made up primarily of families in the lower socioeconomic levels. According to the plan of the center, physicians, who were part-time employees, usually spending two hours a day seeing patients, were expected to inculcate patients with the principles of hygiene and healthful living during the period of the visit. For the most part, however, physicians saw their role as that of "control," routine checking of health and administration of curative services as needed.

Nurses were expected to supplement the preventive medicine of the doctors and carry out health education. This they did by seeing patients after the doctor had seen them, explaining and reinforcing his instructions, and by visiting patients in their homes. Since physicians were of high social status and patients of low status, poor communication frequently existed; the nurse's amplification of the doctor's instructions was therefore a very important part of the patient's visit to the center. Furthermore, regular clients were reassured when upon visiting the health center they would find their friend, the nurse who visited them in their homes.

But when the director was unable to obtain the services of regular health educators to carry out an enlarged health education program, he decided to accomplish this end by relieving his nurses of clinical duty and assigning them to full-time home visiting, leaving only one nurse for the irreducible minimum of clinical duties. The director thought of the change as primarily an administrative shift and a "technical modification" in the nurses' duties. Nevertheless, it produced significant changes in patterns of relations between patients and center personnel. Health education in the clinics suffered, since the physicians, who were in any case largely uninterested in this aspect of their work, were no longer buttressed by nurses. Contact between doctors and nurses became dependent primarily upon patient records, and this meant that desirable follow-up attentions often were delayed, in some instances for months.

In general, the nurses were pleased with the new arrangement; perhaps in part they appreciated being out from under the immediate control of the physicians, who felt the nurses' role was supplementary to their own. On the other hand, patients came to the center with less confidence than before. When they no longer found on duty the nurse who was their friend, who knew them in their homes, they felt strange and uncertain as to how to proceed. Further, without the intermediary function of the nurse the doctor's instructions were less well understood, his instructions were not always met with the previous degree of confidence, and sometimes patients left without bothering to fill prescriptions.

At the time the study was made it was still too soon to tell whether these losses were balanced by the gains from more intensive home visiting. It is clear, however, that the social structure of the health center was a vital element in the combination of factors that make for an efficient organization, and that what appeared to be a routine administrative change had implications far beyond what anyone had envisaged (Simmons 1955b).

# 11

## the anthropologist at work: stages of analysis

The role of the anthropologist, and of other behavioral scientists, in technological development programs can be examined further in terms of a sequence of events. A technical aid project has an inception, a planning period, an operations period, and, perhaps, a stocktaking or evaluation period in which lessons learned are analyzed for future guidance. The theoretical stand of the anthropologist assigned to such a program is the same at all stages, but the kinds of things he does and the kinds of answers he gives depend on the particular point in the sequence at which he finds himself. For purposes of illustration, the work of the anthropologist at these four stages will be described under the headings of (1) prestudy, (2) planning, (3) ongoing analysis, and (4) evaluation.

### PRESTUDY

When a developmental project is anticipated, data must be gathered so that planning can be done intelligently. The research an anthropologist must do at this point will depend in large measure on the basic cultural research that has previously been carried out in the area concerned. If an anthropologist assigned, let us say, to a public health project already has a good idea of the social structure of the area involved, the economic patterns, the value system, the folk medical beliefs and practices, he can direct himself immediately to the specific subjects that presumably will prove important in the proposed project. His task will be to draw upon the corpus of unspecialized sociocultural data available, decide what specific information is lacking, and then attempt to fill in these gaps. Prestudy, of a few weeks and sometimes less, will then give the answers that are needed.

If, on the other hand, relatively little is known about the culture and society of an area, the anthropologist will need to do much more research in order to give the same degree of help, for he will have to work out the broad patterns of life before he can answer specific questions. This is why basic anthropological research is so very important if efficient use is to be made of anthropologists in goal-oriented programs. What is already known constitutes scientific capital; it forms a plateau of sociocultural knowledge, which permits more rapid work than does a takeoff from a sea-level plain. A behavioral scientist can work much more effectively in a technical aid program in, say, Mexico or India than in one in Afghanistan or Nepal simply because so much basic research has already been done in the former countries, whereas in the latter countries very little has been done. Generalized basic research—research not directed toward specific problems—also is very important, because, until work is begun on a developmental project, it is not possible to recognize all the significant factors.

The necessity of encouraging and supporting basic behavioral science research is, unfortunately, something that most government officials understand only with difficulty. They want specific answers to specific questions—and quickly. Their position is not hard to understand; they are prisoners of the annual budget. Results must be shown, to justify continued support, and funds earmarked for work that may not pay off for several years seem a less attractive investment than those that may show quick returns. Still, if the best behavioral science support is to be given to technical aid programs, much more basic research must be sponsored by the organizations that administer such activities.

It should be clear, then, that the nature of anthropological prestudy will depend on the amount of previous anthropological research. If little has been done, a prestudy will, by necessity, be no different from basic research. If much has been done, a prestudy will focus almost immediately on the goals of the proposed project. The work of Isabel Kelly and her colleagues, described in the last chapter, stands midway between these two poles. Initially she believed that housing in El Cuije could be studied to the near exclusion of other aspects of culture, because of what was already known. This turned out to be true only in part, since El Cuije is in a subcultural area of Mexico that had been largely overlooked in earlier studies. Consequently, the seven-month study embraced farming techniques, family budgets, work patterns within the family, political factors, and many other items that proved significant to the immediate task at hand. At the same time, from the very beginning, this research was specifically pointed toward housing, and little attention was given to such things as folklore, music, religious practices, and death observance, which, however interesting, had little to do with project goals.

A prestudy of a different type is described briefly by Barnett. Woleai, a group of 23 islets ringing a large lagoon midway between Truk and Palau, was used as a Japanese air base during World War II. Coconut and breadfruit trees were cut down to make room for military installations, the women were evacuated to other islands, and the men were put to work. At the end of the war the islets were largely denuded, much valuable land had been covered with coral and concrete, and agricultural possibilities were greatly reduced, compared with the prewar period. As the people returned to the islands, it was clear that the American administration was faced with an assistance and welfare problem.

In 1950 a survey team was sent to study the situation and make recommendations. The anthropologist was to report on social, economic, political, religious, and educational conditions and to determine the inhabitants' needs for assistance. He found that the common hardships experienced by the natives had produced a community with a high degree of integration and that they had shown realism in working out means of food production and distribution. But although they had defined their basic need as more food, it appeared that they were not really facing starvation. The anthropologist therefore recommended against handouts of rice and canned goods, which he felt would only raise their future expectations. He expressed the belief that some cash income—which had existed in Japanese times—was the most pressing need of the islanders, and that this should be obtained by efforts to improve agriculture and introduce other foods that the natives could raise themselves (Barnett 1956:94–95).

## PLANNING

With general background knowledge and, ideally, a prestudy directed toward specific problems, an anthropologist should be able to predict in broad outline the probable consequences of any proposed detail. Conversely, he should be able to explain the minimum preconditions for success in innovation, or point out the unforeseen problems that will constitute barriers. In planning, the anthropologist is doing systems research, making use particularly of his concept of culture as an integrated unit, in which one dislocation or change will affect, and be affected by, a whole series of other factors. In an Indian village the Western technical aid agent may feel that composting of cow dung will go a long way in solving problems of fertilizing fields. The anthropologist, on the other hand, will consider the total situation. He will point out that manure has many uses, all competing for the limited amount available. It is used as fuel in cooking, and its slow-burning characteristics make it particularly important in the preparation of ghee, a clarified butter used in a variety of foods; manure is important in the mud used in house plastering; and a little is even used in the hubble-bubble pipe, around which male social gatherings center. Consequently, less than half of the dung produced by village cows remains available for fertilizer (Marriott 1952:265).

### Planning in a malaria eradication program in Mexico

Although anthropologists have been used more often in planning than in prestudy activities, there are remarkably few examples on record that show exactly what has been done. Isabel Kelly and her colleague, the Mexican anthropologist Hector García Manzanedo, report on one instance. They were asked by the Mexican Ministry of Health and Public Assistance to examine the Mexican Government's antimalarial program—a part of a worldwide attempt to stamp out malaria—and to make recommendations for more efficient operations. This research was done without fieldwork, on the basis of the two anthropologists' wide general knowledge of Mexico. First they examined maps showing the infected areas, and then they superimposed these on maps showing the distribution of the Indian population of the country. This revealed that many of the areas of high infection were also areas of dense Indian population, often with many monolingual groups representing a number of different languages. These discoveries suggested that in such parts of the country there would be far greater problems of communication—and consequently greater costs on a per capita basis—than had been anticipated by the medical planners.

Malarial control requires blood samples, and the anthropologists pointed out that there is widespread reluctance among many Indians, and often among rural mestizos as well, to permit blood to be taken. Sometimes this opposition is based on the belief that the blood can be used for witchcraft directed against the victim through processes of sympathetic magic, in which any evil done to something from the body will react in the body itself. In other areas the opposition stems from the belief that blood is a nonrenewable substance and that to the extent that any is lost a person loses strength and sexual vigor. In such areas the anthropologists suggested that withdrawal of blood possibly might be accomplished only through force. On the other hand, they pointed out, in parts of Yucatán and Quintana Roo, withdrawal for diagnosis is a part of traditional therapy, and in these regions, they suggested, there is apt to be less resistance.

Based on observations where earlier malaria-control work had been done, they knew that insecticides often cause the death of small chicks, bees, and even cats, and that these deaths have aroused much antagonism. They warned of the need to explain very carefully the effect of DDT and to take measures to reduce harmful side effects to a minimum.

They also studied the administrative layout of the project and noted that the thirteen major zones of operation had been established largely on the basis of population. Each zone was to have essentially the same number of workers and the same plan of attack. Some of the zones were relatively homogeneous in population and offered no special problems. But those that included many different Indian groups, often in remote and isolated areas, obviously would require more workers—and workers with special talents for dealing with Indian groups. The anthropologists therefore suggested that the antimalaria organization seek the cooperation of the National Indian Institute, which already had developmental centers in a number of Indian areas. They felt that the "cultural promoters" already at work, who had gained the confidence of the

Indians (and many of whom were Indians themselves), would be invaluable in aiding the antimalaria campaign in these areas (García Manzanedo and Kelly 1955).

## ONGOING ANALYSIS

Potentially, ongoing analysis is the area in which the greatest scientific advances may be made; it should therefore be the most gratifying to the anthropologist. In this situation the anthropologist has, or should have, the opportunity to test his hypotheses by seeing whether his predictions come true. He can see, immediately and firsthand, the consequences of innovation, and hence study, under near-laboratory conditions, the whole process of acceptance or rejection of new elements. He talks with informants, notes their attitudes and their reactions. He can find out, with little effort, just *why* people do or think what they do. Competent technicians sometimes do the same thing. Nevertheless, the average technician rarely has time for extended questioning, even if he knows the techniques and has a mind for it. The anthropologist in this situation ideally should provide the eyes and ears for the project. His sensitivity to developments can make it possible to modify plans while there is still time to change them, and to experiment with new or altered ideas.

The anthropologist is apt to be the first person to spot barriers as they develop, and he should be in a position to suggest ways to overcome them. He is able to suggest experiments with alternate techniques and to measure and evaluate the relative effectiveness of the methods tried. For example, in health education, should the major emphasis be on movies, film strips, or the use of puppets? In communicating new techniques, is radio or television superior to other audiovisual approaches? The best methods in a target area cannot be determined until several have been tried out. The anthropologist (and other behavioral scientists) can set up controlled experiments and give rather good answers to questions like these.

### An experiment in Ecuador

A recent example of such a controlled experiment in Ecuador illustrates this point. It was designed to provide information on the relative effectiveness of several kinds of communication methods— radio, audiovisual communications (including films, slides, posters, lectures, and demonstrations), and a combination of the two media.

The object of the research was to compare the impacts of the three methods in promoting the construction of latrines and smokeless stoves, the making of marmalade, and the acceptance of vaccination. In the test area six villages were selected: Three were controls, with no intervention beyond announcement of the program and the information that supplies and tools were available. In the fourth town radio alone was used; in the fifth, audiovisual alone; and in the sixth, a combination of the two. The initial two-week period was used to inform people that the campaign was under way and to persuade them to participate. The next phase, seven weeks in length, was designed to continue to motivate people to participate, but its prime purpose was to instruct them in the techniques of the recommended practices. At the end of the nine-week campaign changes in behavior were counted, and after three-month and six-month intervals sample checks were carried out to find any additional changes and to determine to what extent the new practices were being continued.

The program was successful in that in the experimental towns significantly more stoves and latrines were constructed and more marmalade was made than in the control towns, and more (but not significantly more) homes were involved in vaccination. Several major lessons were learned:

1. Contrary to expectations, combined radio and audiovisual campaigns were *not* more successful than either radio or audiovisual techniques alone. Since expenditures for the the campaign (per household) were held constant for the three approaches, the conclusion was drawn that communities reach a saturation point beyond which additional communication efforts have no impact.
2. Radio persuaded more people to participate than did audiovisual or mixed radio and audiovisual approaches. In "Radio Town" at least one of the three "active" practices—stove, latrine, or marmalade making—was adopted in 55 percent of all homes, versus 39 percent in "Audiovisual Town."
3. The data on individual activities show that radio was the most effective in promoting adoption of stoves, marmalade, and vaccination, but it was less effective than audio-visual methods in stimulating latrine construction. In Radio Town latrine construction constituted only 14 percent of the three active practices undertaken, compared with 41 percent

in Audiovisual Town. The authors conclude that since stoves and marmalade are made by women, and latrines by men, differential exposure to the media is responsible for the differences. Women spend more time at home, so are exposed more constantly to radio, while men spend more time away from home, with greater potential exposure to audiovisual efforts.

4. The authors conclude also that the apparent superiority of the audiovisual approach for latrine building may indicate that it is in fact the best medium for relatively complicated and expensive projects where more technical details are involved.

5. In addition, the audio-visual approach seems to have had longer lasting effects than radio, in that during the nine-month follow-up period it accounted for a relatively higher proportion of continued activity.

"It is obvious from these findings," say the authors, "that different media have different optimal uses" (Spector et al. 1971:45). Clearly, in a major national campaign, initial experiments such as this can provide effective guidelines for achieving maximum impact at the lowest possible cost.

## A Guatemalan case study

Program administrators often think of anthropologists primarily as troubleshooters. Anthropologists "know the local culture," so when something goes wrong, they can be sent in to pull the fat out of the fire. R. N. Adams tells of an instance in which he was cast in this role in a health program in Guatemala. The Institute of Nutrition of Central America and Panama (INCAP) was carrying out nutritional and health research work in several Indian villages near Guatemala City. Food supplementation was provided for school children as a part of a program to determine how the local diet, which appeared deficient in protein, might be improved. In addition to being fed, children were given periodic physical examinations, which involved X-rays and blood withdrawal. In order to promote the best possible relations with villagers, a Guatemalan social worker was employed to work in homes, and a clinic was established in each village with a full-time nurse, on the assumption (frequently sound) that catering to the health needs of all villagers would win friends for the program. At first the work went well, but then villagers in one community began skipping appointments, rumors circulated that the project was politically oriented, parents said their children were being injured, and hostility reached a point were continued work seemed doubtful.

At this point Adams was asked to analyze the problems. After limited research he decided tentatively that three major problems existed: dissension and, especially, poor communication among INCAP personnel; a disturbed national political situation, which had local repercussions that adversely influenced the work; and the rather extensive social work program, which had been designed to win friends but was, in fact, producing more trouble than aid.

With respect to the first problem, the most interesting thing is that prior to investigation the field team had placed the blame for lack of Indian support on the Indians themselves. Actually, as Adams says, "the trouble lay within the organization of the field team itself, and the Indians were little more than uncomfortable bystanders" (R. N. Adams 1953:11). This is one of the first instances on record in which shortcomings within the responsible bureaucracy itself were identified as a major barrier to change. Since personal dislikes between some of the team members had reached a point where they could no longer effectively communicate with each other, Adams assumed the role of mediator between the disputants until a more congenial group of workers was assembled, in spite of the fact that nominally he was only a researcher.

With respect to political factors, the problem lay in the existence of two major national factions: a progovernment group, thought by many to be communist-led, and an anticommunist group. The Indians, as Catholics, were anticommunist. Early in the program Adams and his co-workers found that they were being called communists, an identification based on the belief that INCAP, actually an international organization, was an organ of the government. The field team solved this problem in a direct and forceful fashion. Its members identified those individuals who were spreading the rumor that INCAP was communist. They then had "fairly strong conversations" with the people concerned, telling them that they had been lying, that they were spreading false information about reputable people, and that by so doing they were actually aiding communism. At the same time the gossip campaign was openly discussed in the homes of those Indians who had been friendly to the village workers. In this fashion the problem was overcome (R. N. Adams 1953:11).

The third problem, that of the undesirable effects of the social work program, seems to run contrary to frequent experience in which one good program bolsters another. This program included bringing a breeding boar to improve local stock, setting up a municipal chicken coop to demonstrate improved poultry husbandry methods, holding social evenings, and other such activities. Unfortunately these activities caused friction between villagers. The schoolteachers felt the boar belonged to them, whereas the local men who had fed it felt they should sell it for profit. The members of the committee that had helped with the municipal chicken coop believed they had the right to the eggs, which they sold for their own profit rather than distributing them among potential chicken raisers. And often the social-evening movies would not arrive or the equipment would break down, leaving the people in an irritable mood. The clinic, too, caused problems; the doctor's visits were poorly timed or unpredictable, and services did not correspond to people's expectations. With the withdrawal of the social welfare program and the reduction in the clinic's services, sources of irritation were removed, thus facilitating concentration on the nutritional program (R. N. Adams 1953:12).

But still other problems remained, centering around folk medical beliefs. Investigation revealed that opposition to blood withdrawal was based on the belief that blood is nonregenerative, that each person has only so much for an entire life, and that to the extent that it is lost the individual is permanently weakened.

> One informant told the anthropologist that the villagers simply could not understand why doctors who claimed to know how to make people well went around intentionally taking the blood of little children, thus making them weaker. Weakness made one more susceptible to illness, so that blood-taking was the reverse of what doctors should be doing. This informant concluded that doctors could not know very much about making people well (R. N. Adams 1955:447).

When the nature of opposition to blood withdrawal was found, steps could be taken to counteract it. In part this involved determining the minimum amount of blood needed for the test and exercising care to make sure that no more than this was taken. In part it meant taking of only a few samples at a time, so that the psychological impact of a mass bloodletting was avoided. But it also meant capitalizing on folk belief. Since blood was considered to indicate strength or weakness, it followed that the condition of the blood could be used as a measure to determine a person's health and resistance to illness. Accordingly, the workers started to explain how the blood withdrawal in small quantities permitted the doctors to tell about the health of the child and to take necessary steps should the child's blood prove to be sick. This explanation was accepted, and when, after an interval of more than two months, blood was again withdrawn, there was little opposition (R. N. Adams 1955:448).

As this instance illustrates, an anthropologist in the role of a troubleshooter often can be very helpful. But troubleshooting is at best a stopgap measure. More and better prestudy and planning should make it possible to reduce this role vastly.

## EVALUATION

Evaluation is one of the most important types of work done by anthropologists in goal-directed programs. The technique is to study a specific program through its history, examining documents and, when possible, interviewing the people who have participated, in order to extract from the experience lessons that can be fed back into improved planning for future work. The anthropologist often faces a special problem here. Human problems in technological change have not, in the past, always received the attention due them; consequently, many praiseworthy projects have foundered. An ex post facto analysis therefore frequently turns out to be a summary of what went wrong and not of what went right. This puts the anthropologist in the role of the carping critic, which does not endear him to the administrator, who feels—correctly—that the anthropologist has not had to face the practical problems of running a project, and that, if he had, he might be more tolerant. It is unfortunately true that most of the readings in the field of applied anthropology are analyses of failures or partial failures. Yet, from the standpoint of scientific research, failure or success of a particular project is incidental; the important thing is whether lessons can be learned that will spell success in the future. The example of evaluation here summarized illustrates one major attempt to use behavioral scientists in program appraisal and shows, I hope, some of the helpful lessons that emerge from this type of work.

### Health program evaluation in Latin America

Between 1944 and 1952 the Smithsonian Institution sent visiting professors in the behavioral sciences to Latin America to teach in institutions of higher learning and to participate with local scholars and students in making basic studies of rural culture. Sixteen volumes ultimately were published, on many aspects of Latin American life. None of this work was pointed directly toward specific problems, but, with other Latin American research, it provided an excellent jumping-off point for action. This opportunity came in 1951 when the United States Public Health Service asked five Smithsonian professors—four anthropologists and a sociologist —to participate in a major evaluation of the first ten years of public health work carried out by the Institute of Inter-American Affairs in cooperation with host-country health departments (Erasmus 1954; Foster 1952; Oberg and Rios 1955; Simmons 1955a). Health centers, a cornerstone of the ten-year program, were selected as the major focus of research, but hospitals and environmental sanitation also received attention. No standard questionnaires were sent from Washington; each scientist was asked to use his judgment as to how to approach the assignment. The sociologist (Simmons) emphasized such things as problems of social structure, status hierarchies in the medical profession, and conflicts stemming from different role perceptions. One anthropologist with long government experience (Oberg) stressed the administrative problems a United States government bureaucracy has in preparing its personnel for work with host country counterparts (incidentally, in so doing he developed the concept of culture shock). A second anthropologist (Kelly), historically and ethnographically oriented, emphasized the nature of folk medicine and folk curing, the rural-urban dichotomy as reflected in behavior patterns, mothers' problems in meeting demands of their husbands, and the like. The other two anthropologists (Erasmus and Foster) concentrated on economic factors and basic patterns of change. Some behavioral scientists would criticize the obvious lack of research design. But this initially flexible approach outlined the parameters of the problem. Preliminary findings were exchanged, basic problems were defined, and agreement was reached on how comparative data from seven countries could be obtained.

Two major problem areas were identified: (1) the quality and nature of interpersonal relations, particularly between patients and public health personnel, but also among public health personnel themselves; (2) the whole complex of beliefs, attitudes, and practices associated with health, disease, prevention of disease, and curing—in the broadest sense, "folk medicine."

It was found, for example, that health center patients often felt a lack of tact and diplomacy on the part of medical personnel.

Part of the problem stemmed from conscious ideas of class and status that are widespread in Latin America, where people below one's station in life are thought not to merit the same consideration shown toward one's equals or superiors. But in other instances apparent rudeness was completely unconscious and resulted from the desire of a nurse or health educator to do a thoroughly professional (and impersonal) job, as illustrated by the experiences of the Latin American public health nurse (see p. 187). The "proper" role behavior of health center personnel, when executed, created in the patients a feeling of coldness and lack of sympathy. Again, patients complained because of very long waits in health centers, and because visiting hours were scheduled to meet the convenience of doctors and nurses rather than the needs of the client group. Finally, mothers often were antagonized because health centers would not accept sick children or would take sick children only if they had previously been enrolled. In developing countries low-income mothers are not particularly aware of the distinction between curative and preventive medicine—a distinction that is a function of the needs, as well as the vested interests of professionals, in industrialized countries. Consequently, when health centers, which the mothers had been told existed to improve health conditions, refused to help them in their hour of need, they often took a dim view of the preventive services whose goals were foreign to their ways of thinking.

This research was among the first to highlight the fact that a successful American public health project cannot simply be transplanted to another country with different health levels, economic potential, and population groups. In Chile the decision had been made to accept all sick children brought to a health center. This won the confidence and approval of mothers; then, with their felt needs taken care of, they often were willing to accept preventive services even though they did not fully appreciate their nature. On the other hand, Mexico City health centers, which initially were operated in a strict fashion, turned away sick children not previously registered. Compared with Chile, those centers showed a high percentage of patient dropouts, and preventive work was much less successful. The lesson learned, and now generally accepted, is that in developing countries public curative medical programs are necessary to win the confidence and goodwill that make preventive programs successful.

One of the significant findings from folk medical research—now documented in many other parts of the world—is that peoples with little exposure to modern medicine dichotomize types of illness: There are those that a physician obviously can treat, such as pneumonia, yaws, malaria, and others that yield quickly to wonder drugs; and there are those that the physician cannot treat, because either he does not know about them or he denies that they exist. These often are folk-defined illnesses believed to be caused by magical or other forces: "the evil eye," "bad air," "fright," and the like. The help of the physician is sought in the first instance but not usually in the second. After all, if a father knows his child has been "eyed" and the doctor says there is no such thing, it is asking a lot for him to believe that the right medical aid has been obtained.

Research in folk medical beliefs highlighted a basic problem facing all medical personnel who work in developing countries or in areas where many of their clients have had limited exposure to modern medical practice. That is, to what extent, if any, should medical personnel cater to folk medical belief? Should all medical practice be carried out in the clinical terms that characterize an American city, or should changes be made in practices that violate folk belief, however superstitious such belief may be? Here are examples of the kind of problems that are met.

Peasant peoples in much of the world believe that a ritual disposal of the placenta is essential for the well-being of mother and child. Frequently this means burying it under the home hearthstone. We know that expectant mothers and their families often resist hospitalization because of fear that harm will come if the placenta is otherwise handled. Do we simply assume that people will have to learn that their views are superstitious, or do we make arrangements to deliver the placenta to the family for traditional disposal? The latter solution is repugnant to some medical people, yet it has been successfully used on many occasions. In prenatal health centers with delivery facilities in northwestern Argentina, initial reaction to hospital delivery was antagonistic. But after the practice was adopted of giving the family the placenta for traditional disposal, attendance increased 20 to 50 percent in several centers. Carlos Canitrot believes this decision was instrumental in increased use of services (communicated by Dr. Carlos Canitrot). In northern Nigeria it is believed that the family must have, not only the placenta, but all blood lost during delivery as well. In order to attract patients for hospital delivery, sheets and bedding must be washed and the water must be given the family along with the placenta (communicated by Dr. Adeniji Adeniyi-Jones). In El Salvador and other parts of Latin America, country people share an old Spanish belief: Delivery takes place more easily if the mother wears, or has under her bed, her husband's hat, or if she wears his shirt or jacket. If catering to this superstition induces more expectant mothers to avail themselves of hospital services, should permission be given to the mother to do what she wants?

There is no easy answer to these questions. Many medical people working under such conditions increasingly believe that we can break out of our own medical folklore and superstition and cast at least some of our services in terms of local cultures. Harmless herbal teas, for example, may make it possible to persuade mothers to give water to infants with diarrhea when they would not give it if instructions simply called for lots of boiled water. If a group believes that three (or four) is a lucky number, medicine might be prescribed every three (or four) hours, or three or four times a day, or even in small units of three or four. If a hospital diet is in conflict with folk beliefs about the proper food for certain conditions, modifications might be made without injuring the dignity of medicine; postpartum diets usually are strictly prescribed in peasant societies, and fear that they will be required to eat foods they know are harmful to them sometimes prevents otherwise willing women from seeking hospitalization for delivery. If research reveals that orange juice is proscribed but tomato juice is permissible, certainly it would seem that tomato juice should be served.

There is no single answer to how far one should go in catering to folk beliefs. In 1957 lightning struck a palm tree in the yard of a tuberculosis sanitorium in Tucson. Because of the significance of lightning in their religion, Navaho Indian patients were greatly upset, and two left the sanitorium. The culturally sensitive hospital administrator brought a Navaho medicine man with his ceremonial paraphernalia to carry out the "sing" that normally would be done on the reservation to counteract the danger brought by the lightning. The medicine man's words and songs were piped to all wards on the intercom, so that each Navaho heard the ceremony and his own personal blessing. After this there was no further talk of leaving the hospital, and a health crisis was averted.

But there is also evidence on the other side. As a part of the Smithsonian research, Charles Erasmus interviewed a number of mothers in a fine new maternity hospital in Quito, Ecuador, that had had immense success after a very short time. The mothers all criticized the hospital for forcing upon them things that conflicted with their beliefs: dangerous food, open windows admitting fresh air, daily baths, compulsory fingernail cleaning, and a host of other routine acts, which they insisted endangered their and their infants' health. But when Dr. Erasmus asked why they came to the hospital, they replied that they had noticed babies born in the hospital were much healthier than those born at home—so culture appeared, in this case, to fly out the window as a barrier to acceptance of new medical practices (communicated by Dr. Charles J. Erasmus).

This particular example, as well as other findings of this Latin American evaluation project, confirmed what agricultural extension agents learned long ago: A striking demonstration is one of the most effective ways to change behavior. This is not too difficult in agriculture. If part of a field is planted and cultivated according to tested scientific procedures, the advantages usually are obvious in a few months. But there is no comparable way to demonstrate the absence of smallpox, whooping cough, or diphtheria, which peasant mothers are quite willing to believe may not come anyway. Educated people interpret the statistics and can be convinced, but this type of logic is not widespread in the world. The fact is that preventive medicine's values are almost impossible to demonstrate within the limits of preventive medicine practice itself. But much curative medicine gives immediate and convincing results. A man who sees leg ulcers caused by yaws disappear by magic after one or two injections of penicillin knows that a physician can do wonderful things.

One of the things we have learned in evaluation studies is that preventive medicine can capitalize on the dramatic successes of curative medicine. When participating in the evaluation of the Latin American health program, I was struck to see, in the city of Temuco, Chile, that there was widespread cooperation from mothers in a BCG antituberculosis vaccination campaign. Six months earlier a serious whooping cough epidemic had threatened. Vaccine was flown in, children were vaccinated, and the threatened epidemic was quickly cut short. Mothers told me that this showed them that doctors knew what they were doing when they said vaccinations would prevent illness, so when the doctors asked their help in the BCG campaign, they were quite happy to cooperate.

Many major technological development programs, such as India's Community Development Programme, now make use of behavioral scientists in evaluation work, in order to feed knowledge back to planning to permit increasingly successful projects. But much remains to be done. One of the most serious shortcomings of evaluation work is that almost no comparative, cross-cultural analytical work has been undertaken. A wealth of experience lies buried in official reports and hidden in the minds of workers who have neither the time nor the training to write up their knowledge for the use of others. It sometimes seems as if the sociocultural and psychological dimensions of planned change must be discovered, and their rules worked out independently, in every new project. We have very few mechanisms for making available to new personnel the accumulated wisdom of earlier programs. An interesting exception to this rule follows.

## The antihookworm campaign in Ceylon

From 1916 to 1922 the Rockefeller Foundation carried out an antihookworm campaign in Ceylon. The history of this pioneer venture in cross-cultural technical aid has been superbly described by Philips (1955). To me the most fascinating, and disheartening, thing about this program is that almost all the problems that later public health projects faced were encountered in Ceylon. The social and economic implications of technical aid were discovered, and many of the correct answers were worked out. Yet until Philips searched the records and talked with participants, no major effort appears to have been made to make this experience available to all health workers. Here are the kinds of things that were learned in Ceylon over fifty years ago.

Barriers to change exist in the suspicion of government, rumors, the fact of free services, the low economic margin that makes it difficult for a laborer to lose a day or two of work, folk medical beliefs in conflict with Western medicine, and differing concepts of the role of the curer. It was learned that preventive services were little understood by people who had pressing curative needs, and the importance of a broad medical (and social and economic) program became apparent when people were irritated by having their hookworm treated but not their leg ulcers and other more painful ailments. The common cultural misconceptions held by technical workers were discovered when they applied their values in an exotic society, and the Americans experienced shock and annoyance at the lack of gratitude that greeted their humanitarian efforts.

One doctor, for example, identified himself with the planters, and became so involved in factionalism and the struggle for prestige that he finally resigned. Some of the field directors tried to express the American idea of democracy in an outgoing, warm-hearted man-to-man approach, which was misunderstood by people who were used to, and expected, a degree of authoritarian treatment from educated people. The Americans learned that, regardless of how friendly they were, in some degree they always were classed with the European masters. It took time for them to learn that dignified, and to some extent authoritarian, behavior was expected of them.

The lessons of the integrated nature of society and culture and the processes of sociocultural change were hard to learn. The program planners intended to make a vivid demonstration of the scientific method by eradicating hookworm in a small area and thus setting an irresistible example to the people to work cooperatively in solving other health problems. Not until near the end of the program was it officially recognized that hookworm control had to be intimately related to other health work, and to social conditions as well. When the Rockefeller scientists realized they could not eradicate hookworm in a given area because of all the factors that converged on any single illness, they shifted their emphasis to a permanent government control program and to the fundamentals of rural health work in general.

The subsequent work of the Rockefeller program was guided by the practical lessons learned in Ceylon, and new policies were adapted to a growing realization of the interrelation of health and social and economic problems. But, until Philips made her study, this invaluable experience was not generally available as a guide for contemporary health workers; it was known to only relatively few people. Similarly, the equally valuable experiences of today's technical experts are, for the most part, going unexploited because of lack of interest and the absence of a mechanism to collect, appraise, and set forth in usable form the lessons of contemporary international technical assistance work.

# 12

## technical aid and behavioral science: some problems of teamwork

If, as I strongly believe, behavioral scientists, and particularly anthropologists, are to play a more active role in technical aid programs, what should their role be? Two main problems arise: (1) the definition of the kind of work an anthropologist should do, and (2) the nature of the administrative relationship between the anthropologist and the action team. Regrettably, except in rare instances no good answers have been worked out for either problem. The history of anthropological participation in developmental programs is often one of frustration, misunderstanding, and lack of good communication between administrators and scientists. On comparatively few occasions has the anthropologist felt completely satisfied with the way in which his role has been envisaged by administration and with the types of research he has been asked to do. On equally few occasions has the administrator felt that he has had the kind of support and aid in solving his problems that he had hoped for from the anthropologist. This has been true in spite of many honest and serious attempts to reconcile differences and to work together.

There is no easy answer to the problems of teamwork, but if the causes of difficulty are better understood, then more effective cooperation should result. The basic problem, it seems to me, arises from the fact that the anthropologist and the administrator are members of distinct subcultures with very different values, premises, and goals. We can properly speak of the "cultural chasm" (to use Saunders's felicitous phrase) between the two fields, just as we can speak of the cultural chasm as a barrier to efficient inter-action between members of truly distinct cultures. The aims, the methods of work, the goals, and the reward system of the admin-istrator are vastly different from those of the anthropologist; con-sequently, to the extent that either—in a working relationship with the other—is forced out of his comfortable subcultural mold, he will feel frustrated and dissatisfied with the arrangement.

### Disciplines and professions

The nature of the cultural chasm separating the anthropologist and his fellow behavioral scientists from administrators and tech-nical specialists can perhaps be seen more clearly if we think of the former as members of a *discipline* and the latter as members of a *profession*.

The assumption underlying academic disciplines is that the search for knowledge, without regard to immediate practical utility, represents the highest value. That this is so is clearly seen in the reward and prestige systems of the scientific and scholarly com-munities, where the highest honors are bestowed on those who make basic or theoretical contributions to knowledge rather than on those whose efforts have been directed to application. Tradi-tionally, scientists have tried to carry out their work objectively, to reduce to the lowest level possible the influence of their personal biases and values. Although behavioral scientists recognize that complete objectivity is not possible—all are creatures of their socioeconomic systems and the premises and values that underlie those systems—most of them believe the interests of both science and mankind are best served when they strive consciously to eliminate value judgments from their work.

Although this view is now questioned by some students and a few academicians who argue that research should be more im-mediately relevent to contemporary social problems, it is still held by most scientists. When we observe how scientists work, we are struck by the fact that they are driven by an insatiable curiosity: They want to find out, to know, to order knowledge in meaningful patterns, to build theories that explain complex phenomena. They are not concerned with the immediate practical utility of their discoveries, although, as we shall see in Chapter 13, they are increasingly preoccupied with the fear that their research findings may, in unscrupulous or insensitive hands, be used in manners antithetical to human well-being.

In contrast, professional work is goal-oriented, and it is based upon an explicit value judgment: There are problems affecting the well-being of man that cry out for solutions. The existence of, let us say, a public health organization means that health problems have been defined, that it has been deemed possible and desirable to solve these problems, and that a bureaucracy has been created to work toward their solution. Science is a major component in any health program, but it tends to be of the applied, rather than the theoretical, type.

In other words, an academic discipline stresses theoretical research, whereas a profession stresses goal-directed action. The two aims are by no means mutually exclusive, but they are different, and when the two aims are pursued in a common project, a reconciliation of interests is essential. Many years ago Sir Philip Mitchell, former governor of Kenya and an early enthusiast for the use of anthropology in colonial administration, humorously yet seriously pointed out the very different approaches usually taken by the anthropologist and "the practical man," the administrator.

> If an inhabitant of a South Sea Island feels obliged
> on some ceremonial occasion to eat his grandmother,
> the anthropologist is attracted to examine and explain the
> ancient custom which caused him to do so: the practical man,
> on the other hand, tends to take more interest in the
> grandmother. The one calls it aviophagy and the other murder:
> it depends on the point of view. Nevertheless, though he is
> regretfully obliged to hang the murderer the practical man
> would much prefer to eliminate his motives, and obviously
> to do that he must know what they are and this the
> anthropologist can tell him (Mitchell 1930:217).

Another important difference in the two fields has to do with ego gratification. In neither field, as pointed out in Chapter 9, is money the principal source of satisfaction, for the monetary rewards of both behavioral scientists and technicians are modest. Each looks for basic satisfaction in the esteem in which he is held by his disciplinary or professional colleagues, according to the standards

set by each field. Public health personnel, for example, feel gratified when they know that their efforts have raised the level of
health in their jurisdiction and that this success is recognized by
their colleagues. Behavioral scientists, on the other hand, feel
gratified when they feel that they have made new contributions
to basic science, that they have formulated new and sound theories,
and that these contributions are recognized by *their* colleagues.
The distinct way in which professionals and research scientists
achieve status in their fields obviously has an important bearing on
how they view their roles and what they hope to accomplish in
any cooperative program.

Obviously, if either a professional specialist or an anthropologist
is asked to do work that impedes his striving toward recognition
by his peers, he is not going to be a very happy person. An anthropologist on an applied project may delight the professionals with
whom he is working; by satisfying the needs of administration he
becomes a "good anthropologist." But the chances are he is not
satisfying himself, and he fears he will be overlooked by the members of his discipline whose approbation he seeks. To explore this
problem further, let us see how administrators (and technical specialists) and anthropologists conceive their respective roles and
what each expects of the other in any joint work.

Looking at the problem first from the standpoint of the administrator, we see that he is charged with the achievement of
ends, of goals that have been determined and for which an administrative organization has been created. As a good administrator
he will, quite properly, use all the material and human resources
at his disposal that will help achieve these goals. The personal convenience and likes and dislikes of the members of his staff, although something to be considered, should be secondary to his
primary task of getting the job done. Furthermore, he is operating
within a budget, which as a rule must be defended annually, and
continued support for his program depends on evidence of progress.
If an anthropologist can contribute something within this framework,
he will be a useful addition. More precisely, here are the kinds of
things the administrator wants from an anthropologist:

1. Specific data pointed toward concrete problems: Are the
inhabitants of a Pacific atoll well governed and content? How can
we persuade farmers to use improved seed? Why do mothers drop
out of prenatal health clinics? Why won't villagers cooperate for
common ends? The administrator wants answers to these and a
hundred similarly specific questions. He is not interested—or

thinks he is not interested—in the basic cultural patterns of a rural area, because these are not specific. The problems most administrators see are of a day-to-day operational nature. When R. N. Adams was first sent to a Guatemalan village (Chapter 11) it was simply to find out why a nutritional research project on school children was arousing parental hostility. No thought had been given to basic social and cultural analysis preceding the initiation of the project so that wise planning might have prevented the hostility. The administrator's definition of his cultural needs means that, most frequently, he thinks of the anthropologist as a troubleshooter: someone who ought to be able to dredge up, from his encyclopedic knowledge of queer customs, an immediate answer to a puzzling and threatening problem.

2. Even when the administrator appreciates the cultural implications of developmental work, he needs pertinent information in readily digestible form. That is, reports must be written simply, with a minimum of sociological jargon. Lord Hailey, writing about the role of anthropology in colonial government, expressed the administrator's plight when he pled for simpler writing in social science reports.

> *Every science must, of course, have its own terminology,*
> *and it will always present some difficulty to the layman.*
> *But I sometimes feel that the work of anthropologists would*
> *be more readily appreciated if they showed greater solicitude*
> *for the weaker brethren who desire to profit by their*
> *research, but are not equipped to deal with the more*
> *esoteric of their terminologies (Hailey 1944:15).*

Lord Hailey was writing at a time when social science terminology was in its infancy; what would he say about the incredible proliferation of complex terms in recent years?

Not only must reports be written in simple and clear language, but they must be as brief as possible so that the administrator does not have to wade through mountains of incidental data, which, however fascinating, do not have the answers for which he searches. Even if he had the inclination, the average administrator does not have the time to read all the anthropological information that may be available on the area in his charge. Sir Philip Mitchell has pungently made this point:

> *There was, especially during the nineteen twenties*
> *and thirties, a spate of special reports and investigations; at*

*one time, indeed, anthropologists, asserting that they only were gifted with understanding, busied themselves with enthusiasm about all the minutae of obscure tribal and personal practices, especially if they were agreeably associated with sex or flavoured with obscenity. There resulted a large number of painstaking and often accurate records of interesting habits and practices, of such length that no one had time to read them and often, in any case, irrelevant, by the time they became available, to the day to day business of Government (Mitchell 1951:57).*

This common shortcoming of traditional reports (from the point of view of action programs) is also recognized by some anthropologists. Forde, for example, writes that many of the African anthropological studies, which should have been of considerable value in framing and implementing policy, were not appreciated by administrators, since they were embodied in lengthy studies or specialist papers.

*While from a scientific point of view these represented a great advance in the functional analysis of native institutions, they often assumed a knowledge of, and primary interest in, theoretical problems, and the relevance of their results to the immediate and even long-term problems of administration was not always brought home (Forde 1953:850).*

3. The administrator needs information promptly. He cannot wait to get information from traditional anthropological channels of publication. Even short articles rarely appear in less than two years from the time the fieldwork is completed, and full monographs often are delayed ten or more years. Given the nature of his assignment, it is unreasonable to expect an administrator to devote much attention, not to say money, to research that at best will be of help to his successor or successor's successor. He must have information promptly. Often this means verbal reports, perhaps at weekly staff meetings; and it certainly means periodic memoranda, which may serve as a basis of justification to his superiors for some of the decisions he makes. Clearly the confidential nature of many of these memoranda sometimes precludes publication.

The administrator has other, more general, conditions that must be fulfilled if he is to be happy in his relation with the anthropologist. He must believe that the anthropologist knows something about the nature of administration and the inherent limitations and handicaps under which any program functions—which limit the course and degree of action that the administrator can take. He must feel that the anthropologist is sympathetic to the goals of his program as well as to the people toward whom a program is directed. The administrator cannot be expected to risk his reputation by accepting the evaluation of the anthropologist unless the anthropologist, too, is in some degree involved.

For the anthropologist also there are certain conditions that must be fulfilled in some degree if his work in the applied field is to be satisfying to him.

1. Wherever the limits may have been set, he must feel that he is allowed to make a study that is technically sound and professionally respectable. This does not necessarily mean that he insists on a full year to make a basic ethnographic study, although the general anthropological complaint certainly is lack of time in which to do adequate work. It does mean working under a policy in which the anthropologist is not at the beck and call of the administrator, to be pulled off one project and set to work on another because a more recent and hence more urgent crisis has arisen. And this is a great danger, given the administrator's frequent concept of the anthropologist as troubleshooter. In other words, if the anthropologist is to work well, he must be a great deal more than an answer-box. He must feel he is doing good and original work in which he can take pride.

2. The anthropologist must feel that a reasonable part of his field research will become available to his profession and be read by his colleagues. He needs the approbation of other anthropologists, their recognition that he is contributing to the field, if he is to be a first-class anthropologist. It is not enough to have the satisfaction of knowing that his administrative superior is pleased with his work. The administrator achieves distinction by success in reaching the goals of his organization; the anthropologist achieves distinction by a creditable list of scientifically sound publications.

3. The consequences of (1) and (2) are that, on other than short-term consultation jobs, the anthropologist must feel that the administrator knows something about the organization of his discipline, its values, and its goals, and that he has the sympathy and support of the administrator in achieving personal ends, within the framework of bureaucratic limitations. Practically, this means that the administrator must take at least a middle-range view of anthropology; he

must resist the temptation to assign the anthropologist to a new task each week, and he must ask himself whether a series of quick answers are in fact of more value to him than fewer but more profound analyses. It means also—and this is the most difficult of all things to achieve—that it must be recognized administratively that writing, not only of reports, but of articles and even monographs, is a major part of any job to an anthropologist. Further, it must be recognized that some of the things that will be written by the anthropologist are not immediately applicable to the ends of the program.

## Scientific capital

In my experience one of the greatest problems anthropologists have had in action programs is to persuade administrators (in the widest sense, government) of the importance of what can be called "scientific capital," the accumulated theory and fact that build up painstakingly over a period of many years. Anthropologists often are able able to give quick, and remarkably accurate, answers because they can fall back on previously accumulated scientific capital; they can build on what is already known. Scientific capital is not something that is built up to an optimum level and then forgotten. Conditions change, new problems arise, new interests develop; so the corpus of theory and fact must be continuously replenished if the needs of both theoretical anthropology and practical administration are to be met. The gain for both is long range, but it is just as real for the latter as for the former. That is, ultimately, the justification of government support for basic research.

In view of the conflicting values and aims of administration and anthropology, and in view of the great variety in types of programs in which anthropologists may be used, it is clear that there is no one ideal administrative relationship. There are, however, certain minimum conditions that must be sought.

1. Both administrator and anthropologist must have a clear understanding of the philosophy underlying the work of the other and of the values and goals attached thereto. Not only must there be understanding, but there must also be acknowledgment of the importance and legitimacy of these values and goals.

2. There must be agreement on, and understanding of, the operational roles of both administrators and anthropologists in any program in which the latter are used. Each must understand the inherent limitations in the work of the other and expect neither special favors nor miracles.

3. The administrator must realize that he, his technical colleagues, and the whole sociocultural system within which his program is carried out are just as legitimate and necessary a research object for the anthropologist as is the analysis of the culture and society of the people toward whom the program is directed. Although a full-scale analysis of the innovating organization is not likely to be undertaken, the program director must recognize that the anthropologist will need to ask him and his colleagues searching questions about how they envisage their roles and goals, and how they go about fulfilling these roles and reaching these goals. Whatever the administrative relationship between anthropologist and technical assistance bureaucracy, the best work is done when the dual nature of the anthropologist's research target is accepted: client group *and* innovating group.

### Working relationships

The most effective relationship in a specific situation will depend on the kind of program, its duration, the status level of participants, and many other factors. Anthropologists' views about how the two groups can work together are more diverse (and, I think, less realistic) than those of administrators. At one pole is the anthropologist who believes he should do basic research to be used by administrators as they see fit, but limiting his direct contribution to short-term consultation devoid of specific recommendations. At the opposite pole is the "direct-hire" anthropologist, a fully integrated member of an action team, and, in unusual cases, an administrator. Between these two extremes one finds a variety of relationships, generally short- or medium-term consultantships, in which the anthropologist is on loan from his university and paid a professional fee by his temporary employer.

The first position is attractive to the anthropologist for several reasons. He is his own boss, he does not have to worry about policy, he can be strictly impartial (although in fact he is usually pretty strict in judging the technicians in the area in which he works), and he can define his own problems. This kind of work has only one drawback, when judged for its practical utility: It rarely has any effect whatsoever until it is interpreted and utilized by other anthropologists. The instance of the Smithsonian Institution's Institute of Social Anthropology is illustrative. The United States government supported wide-scale basic research in Latin American society and culture, ostensibly for practical purposes. But the work—16 volumes in all—

was completely unknown to, and doubtless unusuable by, those organizations that properly should have used it; not until the anthropologists themselves participated in a public health evaluation project were the data of practical—as distinguished from scientific—value (Foster 1967b).

In contrast to this situation, the long-term direct-hire anthropologist not only studies the people toward whom a program is directed, but to some extent, and unavoidably, he also studies the organization to which he belongs and the relationship between the two systems. He supplies technical information, and at times he implements it. For many anthropologists this relationship is abhorrent. It partakes of social tinkering, and the anthropologist, enmeshed in everyday practical problems, is thought lost to his true calling. Wilson states this position well when he says

> The conception of "technical information" . . . is the key to
> the correct relationship between social scientists . . .
> and men of affairs. For human societies . . . have a hard reality
> which cannot be mastered without patience and objective
> study. It is the scientists' business to undertake that
> patient and objective study, it is the business of government
> and industry to make use of their results in the fashioning
> out of the present whatever future they desire (Wilson 1940:46).

Largely because of the extent to which this view is shared by other leading anthropologists, applied work is often viewed as second-class anthropology, without value as far as anthropological theory itself is concerned. Consequently, it is not easy to get the best anthropologists to serve in applied programs until after they have made their reputations in basic research and theory building.

Fortunately this extreme position is by no means universal, and increasingly it is recognized that direct integration in an action team permits an anthropologist opportunities he would not have while working independently. The most important of these rewards is that as a working member of a technical aid mission an anthropologist can more effectively study the innovating organization. The anthropologist takes it for granted that he must gain the confidence of natives among whom he works; it is more difficult for him to see that this is equally true of the members of the bureaucracy. When he attends regular staff meetings with the others, is subject to the same series of frustrations and delays, is obviously one of the gang—and if, by chance, it is known that at some time in his life he has held administrative responsibility and been faced with budgetary and personnel problems—his problems of establishing rapport are minimized. In this setting the kind of work an anthropologist does is different from that in a basic community study. He spends less time with local peoples and more with the bureaucracy. But then his problem is not just local people: It is a pair of systems, studied as they interact.

The sympathy with which an anthropologist may view direct hire or maximum integration with a bureaucracy will depend on his experiences, and these are many. My own experiences with administrators have been generally agreeable as well as profitable; perhaps for this reason I lean more toward close working relationships than do many anthropologists. But there are others who also recognize the limitations in the minimum integration conception. Barnett, speaking of a variety of reasons why anthropologists and administrators do not make contact, writes that "the essential difficulty in all these instances is that the research specialist is regarded as a stranger, often as an interloper, by regular government officials. He has no status within the organization, so his views can be treated like those of any other outside observer or critic" (Barnett 1956:172). When an anthropologist is a member of a team, by job definition he is expected to express opinions. He may be disregarded, but he is even more sure to be disregarded if he is a technical adviser who operates entirely on his own initiative, independently deciding what should be explored, and not being informed about, or appealed to, on matters that the administration regards as problems.

In Department of the Interior policy in the Micronesian Trust Territory, direct representation of the social scientist's view on policy matters at the top level is regarded as essential, "and where it has not been looked upon with favor by a departmental officer, the absence of a consultative pattern is taken to be a major obstacle to the success of the experiment (Barnett 1956:173). Barnett speaks of the ineffectiveness of the "free-lance anthropologist or the administratively isolated technical specialist," as well as of advisers and consultants who suffer from lack of familiarity with administrative processes and their day-to-day demands.

> No one outside an activity can have the acquaintance with
> its requirements that is a prerequisite for the preparation of
> readily adaptable advice. . . . Policy determinants and
> operational rules may not be appreciated in recommendations
> by outsiders; and an unfamiliarity with internal power

*struggles and personal clashes often destroys the utility of a
report simply because it has been phrased one way rather
than another or presented to the wrong individual (Barnett
1956:174).*

Barnett, it should be noted, is not arguing for an anthropologist-
administrator; in fact, he expressly rejects this in favor of a sharp
demarcation between the two functions. He is pointing out, and
more clearly than it usually has been done, that the best scientific
work may well be done with a close administrative relationship.

There are a few anthropologists, whose theoretical competence
is not in jeopardy, who admit they like to make value judgments
based on their research and to do what they can to implement these
judgments. One of the most convincing statements on this position
is by Thomas Gladwin, who first did basic research under U.S. Navy
sponsorship (Coordinated Investigation of Micronesian Anthropology, CIMA) on Truk, then served as anthropologist on the Civil Administration Unit, and finally took the position of Native Affairs
Officer, in which capacity he was responsible for the political and
economic affairs of the 15,000 natives of Truk and surrounding islands. While doing basic research Gladwin learned the language and
acquired an unparalleled knowledge of the local culture, which he
described in a monograph that has achieved the status of classic in
the field of culture and personality. Thus, he was well equipped to
make decisions. Further, he was faced with no major ethical problem: The basically altruistic U.S. policy toward Micronesia was one
to which he could wholeheartedly subscribe.

Gladwin obviously took satisfaction in feeling that he was directly helping the Trukese in his role as administrator. "It is my thesis,
perhaps in self-justification, that, provided he has the opportunity to
become adequately knowledgeable in the local culture and can develop adequate rapport, the anthropologist who wants to be useful
in administration must himself become in some degree an administrator" (Gladwin 1956:64). Gladwin found that ultimately he came to
the point where he required another anthropologist to do most of
the original field research, and his research subsequent to becoming
an administrator was done on the run. But this he felt was compensated for by his unique competence to mediate between high-level
policy and the immediate situation.

*The anthropologist, equally sensitive to American and to
Micronesian values, can at least hope to achieve a balance*

*between them which will do violence to neither. He is not only best qualified to do this, but also when he implements his own convictions he is, in the best American tradition, sticking his own neck out. If he fails, his failure will be weighed against his successes. But if someone else takes the responsibility for implementing the anthropologist's recommendations and they produce a fiasco, the result can be loss of faith in the anthropologist and a period, at least, of sharply reduced effectiveness. If the job is worth doing, the anthropologist must do it himself (Gladwin 1956:64–65).*

My position is close to that of Gladwin. Decisions are going to be made that involve anthropological knowledge. Without quibbling as to whether he is acting as a citizen or a scientist in so doing, I think the anthropologist has a clear duty to society to participate in decisions in which, by training and experience, he is obviously an authority.

I would beg the question as to what is the best administrative relationship between anthropologist and technical aid program, and suggest that the task is to make sure that the representatives of both fields understand and appreciate the nature of the other's values and methods of work, and that for a specific assignment there be agreement on respective roles. A particular administrative arrangement then becomes secondary. In most cases a rather close relationship will be the answer, but this need not be a *sine qua non* for mutually profitable results.

# 13

## ethics in planned change

Deeply embedded in the American conscience is the belief that those who can should help their less fortunate neighbors. In response to this value Americans have established and given generously to community chests, united crusades, and myriad other voluntary associations in the fields of health, education, social welfare, and race relations. The same urge has made itself felt on the international level, from the time of nineteenth-century missionary efforts to the contemporary activities of government and private organizations in health, agriculture, education, and other "nation building" programs.

So much a way of life is aid to others that we rarely ask ourselves, "Why are we doing this, and what are the ethical implications of our actions?" To give, to help: For most of us, that has been all the justification that is necessary. Yet very genuine moral and ethical problems arise in every instance in which attempts are made to change the way of living of others. Great good may, and often has, resulted from American efforts to help others. Our technical know-how in plant genetics and agronomy has played a major role in finally bringing to fruition, after a generation of uphill struggle, the Green Revolution, a quantum increase in agricultural production in developing countries. Yet, as we have seen, any major change produces consequences that may be good, bad, or both, for recipient peoples. The successes in recent years in environmental sanitation, public health, and medical services, which have drastically lowered death rates and extended life spans, illustrate well the dangers in a single-minded professional approach not linked to the wider strategies of development. Public health and other medical workers justify their activities on the grounds that "health is better than sickness." It is, of course, difficult to quarrel with this cliché, yet failure until very recently to integrate birth control with death control has produced a population problem far more threatening to man's future than unchecked disease. In spite of the success of the Green Revolution, which has bought us a few years of grace, it is still legitimate to ask, "Will four billion undernourished people be more desirable than two billion undernourished people?"

In this final chapter I want to consider three aspects of ethics in planned change. The first has to do with the broad question of American government and private agency technical assistance in developing countries, its rationale and justification. The second treats of the individual ethical posture of the technical specialist who works abroad, as a program planner, administrator, or substantive expert. And the last is concerned with a more narrowly professional matter: the anthropologist, his role in helping to bring about change, his relations with his government, and the ethical guides he follows in his work.

## Ethics in technical assistance

When we look for the rationales underlying major technical assistance efforts, we quickly encounter more than a humanitarian motivation, a desire to help needy people. Nineteenth- and twentieth-century missionaries who established hospitals and schools saw these institutions as aids in saving souls for Christianity, in bringing light to the heathen. In early international medical programs the rationale was the same as that underlying the first public health work in the United States: In a crowding and mobile world, pockets of large areas of disease are a threat to the healthy, so one's own health is best protected by protecting others. I have also suspected that the personal element of ego-gratification may loom larger in major programs for eradicating disease than in other fields. What activities have the potential for greater professional exhilaration and satisfaction than simultaneously gaining the plaudits of one's peers for spectacular medical successes and earning the gratitude of those whose lives are saved or made more healthy?

Turning to the single largest technical aid program in history, that of the Agency for International Development and its predecessors, beginning in 1942 with the Institute of Inter-American Affairs, we find various rationales—all adding up to national self-interest. The early justification for American health, agriculture, and education efforts in Latin America was based on World War II problems. Supplies of natural rubber, which came from Malaysia, were threatened by the Japanese, and the Amazon jungle offered the only feasible alternative source; protecting the health of rubber workers was thus a major argument for early medical programs. Another problem was food. Since most Latin American countries lacked good internal transportation systems, much food was transported from one country to another, or within large countries such as Brazil, by coastal steamers; in addition, many countries were net importers of food. Submarine warfare gravely threatened these normal food supply routes, and thus placed a premium on increasing local yields to make countries and regions internally self-sufficient. Both the health and the agricultural goals could be more easily achieved, it was argued, if people were literate. Hence, programs in education were introduced. Further justifying such efforts was the Good Neighbor Policy formulated by Franklin D. Roosevelt, which had as goal strengthening economic, cultural, and friendship ties between the United States and our usually neglected neighbors to the south. In many of these war programs anthropologists played important roles as technical consultants on local customs and as specialists in the field of cross-cultural communication.

With the extension of wartime Latin American technical aid programs to most of the free world in the early 1950s, the rationale became broader. Most Americans uncritically assume that our country has created in its economic, social, and political systems, the most desirable—in fact, the inevitable—model for "progress," and that sooner or later other countries will follow along in our footsteps. After World War I we viewed communism as a temporary aberration, a threat and a nuisance over which our moral integrity and material well-being would prevail. Rather naïvely we believed that people in all the rest of the world wanted to live like Americans and to be like Americans. Given the memory of the millions of Europeans who had immigrated to the United States, and who *were* anxious to become Americans, this was not an illogical conclusion. Poverty, coupled with poor health, primitive agriculture producing insufficient food, and limited education—these, it was argued, were the conditions that inhibited peoples in most of the rest of the world from making the progress they desired toward the American way of life.

Beginning with our world-wide technical aid programs in the years following World War II, the argument that swayed Congress in appropriating several billion dollars a year—an argument that most Americans genuinely believed—was that developing nations had neither the technical skills nor the financial means to lick poverty, disease, malnutrition, and ignorance. The response seemed simple: Send great numbers of skillful technicians to work with local "counterparts"; bring promising young people to this country for technical educations; and lend money or make outright grants where necessary to build strong countries. With sufficient economic progress, and improvements in the level of living of the masses, we assumed that most countries would remain friendly to the United States, and that they would develop reasonably liberal democratic political systems. We wanted to "win friends and influence people." We hoped that, by so doing, we could create a world in which we could live without fear of military attack, and in which we would have access to those raw materials needed to maintain and further develop our standard of living.

In retrospect, one can only be astonished at our naïvete, for the dynamics of change and development are infinitely more complex than our simple models assumed. Perhaps we were most shortsighted in our failure to realize that technological change is not just a question of giving up old practices and learning new and better ones, of educating and training more and more people. Although in a literal sense cultural change consists in abandoning old ways and adopting new ones, the extent to which this process can be effective in raising living standards is determined and limited by the socioeconomic system within which people live, and especially by the power structure that controls the system. For example, for a good many years most American agronomists have recognized that traditional patterns of land ownership in developing countries—large estates worked by landless peons—severely limit the individual progress that small farmers can make. Some of them have preached land reform and urged foreign governments to recognize the gravity of the situation. A few countries, such as Mexico, have carried out important land reform programs. But governments in developing countries frequently are controlled by those who have most to lose through land redistribution, and such oligarchic establishments have shown little enthusiasm for major changes. Other essential social reforms similarly are impeded by governing elites who are reluctant to give up traditional privileges.

Many knowledgeable Americans, including some with wide experience in the international field, are now asking difficult and disturbing questions: Given the reality of autocratic and often dictatorial governments in many developing countries, and given the reluctance of the wealthy classes to carry out major economic and social reforms, *is technical assistance ethically justified?* Is technical assistance, to the extent that it brings modest improvements in living standards, in fact simply propping up the existing power structures, helping to maintain the status quo by slightly alleviating social discontent? Are most of the world's nations going to develop socioeconomic and political systems that are significantly different from those of the United States, because our model is inappropriate for them? Are we helping to postpone the day of reckoning, meanwhile building up pressures that will make ultimate explosions far more violent?

These and other similar questions are being seriously asked by thoughtful and experienced people. We do not have to go outside the Americas to find evidence suggesting affirmative answers to them: first, revolutionary changes in Cuba, and next a Marxist president in Chile, long thought of as a bulwark of democracy. In other Latin American countries, such as Colombia, Peru, and Bolivia, the masses are demanding land redistribution and other reforms, sometimes—as in Bolivia—taking matters into their own hands. Twenty to thirty years of technical assistance in these countries has had little measurable effect in creating liberal democracies with rising standards of living and greater social equality. On the other hand, Mexico, which experienced its great social revolution more than 50 years ago, in a fortunate time when international intervention was minimal, has made remarkable progress in industry, agriculture, education, and health services. The reasons for this progress, which unfortunately has not reached all Mexicans in equal scale, are varied. One of the most important is that Mexico—through its own universities, and because proximity to the United States has permitted scores of thousands of students to acquire technical and professional educations in this country—now has a corpus of highly skilled basic scientists, medical personnel, agronomists, and educators unparalleled in the developing world. Very little of this technical knowledge can be attributed to direct foreign aid programs; indeed, Mexico requested termination of the American program in the mid-1950s. Yet the essential fact is clear: Technical skills, however acquired, have been basic to Mexico's development.

In assessing the extent to which technical aid may work against the interests of the masses, by propping up oligarchic and undemocratic political systems, we must remember that the governments of new and developing countries do not conform to a single model. It is true that in most of Latin America, economic, and hence political, control has since colonial times been in the hands of a relatively few powerful landowning families, and most Americans familiar with the scene believe that this centralization has worked great injustices on the masses. In contrast, in the new nations of Africa the picture is very different. Here there are no traditional powerful families that have controlled countries for generations, no entrenched establishments. Inevitably new power structures are developing in these countries, and with time they will become increasingly rigid. Yet to argue that we should not extend aid to African nations because we are supporting the entrenched interests is patently absurd.

We cannot, therefore, in blanket fashion write off technical assistance to the developing world simply because in some countries existing power structures may subvert it to their own ends: The need is yet great. But we must be more sensitive to this issue than we have been in the past. In asking the question, "What will be the consequences of this project?" we must go beyond narrow technical considerations and evaluate the impact on the power structure also. Programs that give promise of major benefits to wide sectors of the middle and lower classes, without strengthening oligarchic classes, should be encouraged. Those in which the major benefits will accrue to those already in power should be rejected. It will not require lengthy evaluation of present and past projects to determine into which category a particular kind of program falls.

As developing countries acquire their own skilled technicians through study abroad and local training in new universities, the role of the foreign specialist will change. There will be less and less need for the teacher, the expert in the transmission of elemental technological skills, and greater need for the supporting consultant. The technical and administrative bureaucracies of developing countries often are controlled by old-line civil servants whose professional training is not so good as that of the specialists who serve under them. They may have been the best available when their organizations were set up, but they now feel threatened by younger people whose superior qualifications they are reluctant to admit. In these situations young and highly qualified specialists may be restricted in what they can do, so that the best use is not made of their talents. The outsider, the foreign technical consultant who can validate these young specialists and lend them the prestige of international support, can play a highly useful role in furthering needed work. Increasingly it is the supportive, validating role that justifies a foreign aid program. For these reasons international technical assistance will be with us for a long time, and the individual technical consultant will continue to need to understand the moral and ethical dimensions of his work.

### The technical specialist's ethical responsibilities

What is a reasonable ethical position for a skilled specialist who works to bring change in a country other than his own? What should he know beyond his professional skills, and how should he define his role? A reasonable position, I believe, must begin with introspection and thought. The specialists must ask himself why he wishes to work abroad, what his goals are, how these relate to his ego and his wish to achieve professional recognition. He must also realize the importance of the premises that underlie his behavior—the importance of his professional, bureaucratic, and national assumptions. In a nutshell, he must know himself. A reasonable position for a specialist should also include an eagerness to understand the culture of the people with whom he works, to look for the good elements in these cultures, and to search for the explanations of traditional ways, as well as an effort to restrain excessive missionary zeal, which leads to inability to see or consider alternative solutions to technical problems. The technical specialist must realize that the expression "cultural chasm" is not just a cliché, but a very graphic summary of the differences that separate peoples in different societies and make cross-cultural communication and cooperation difficult.

The most successful technicians, those who are on the most solid ground, are those who begin their work cautiously, who feel their way, who establish local bearings before undertaking major activities. They are the ones who have avoided the blind ethnocentrism that has led some specialists to assume there can be no doubt about the superior merit of their own style of life. To take a particular example, they sense that a bone-crushing, fraternity-style handshake accompanied by a beady stare in the eye of a host-country national does not necessarily connote sincerity. They know that a friendly smile, an extended right hand, a lapel label saying "Hello, my name is John Smith," and unlimited energy to pitch in and work are not enough to justify one's presence in a foreign country. The successful technician is sensitive to the fact that professional aid looks very different to the recipient than to the donor. Most Americans see their presence in a foreign country as reflecting obvious goodwill—an evidence of eagerness to help, to share good things with others, a symbol of generosity and selflessness, a gesture that cannot be misunderstood or misinterpreted.

What does an offer of technical aid imply to potential recipients? It implies many things, such as possible modern imperialism, the danger of economic penetration, a shattering of traditional values, and it means also often-insensitive strangers taking the best houses and driving up the cost of living. Above all it says, in essence, "if you people will learn to do more things the way we do them, you will be better off." This is not a very flattering approach. Consummate sensitivity in proffering aid, mastery of economic, technical, and social facts needed to carry it out, and awareness of the limitations of such aid are absolutely basic to successful cooperative developmental work.

In suggesting an ethic for the American technical specialist working abroad, it is difficult to overemphasize the importance of knowing *why* things are done as they are, before attempting to change them. This is not simply the morality of cultural relativism. It is not simply tolerance and broad-mindedness. It is basic wisdom for the technical agent. In every culture there is a reason for each element. Sometimes the reason is no longer valid; change then will not be dangerous. At other times the reason is still valid, and change may spell disaster. An example that illustrates this most common of errors is the attempt to introduce moldboard plows into areas where they are destructive to soils or are otherwise unsuitable. English has described this problem in Iran:

> Not all the tools of modern agriculture can be integrated
> into traditional Iranian agriculture successfully: the moldboard
> plow is an example of an innovation that failed; it has
> nowhere replaced the oft-maligned nail plow, which has
> distinct advantages over its more modern counterpart. The light,
> wooden nail plow can maneuver in small irregular plots
> bounded by irrigation ditches. It leaves no open furrow
> at one side of the field, which would create irrigation problems
> by collecting too much precious water. The nail plow is
> inexpensive to construct, can be built entirely from local
> materials and requires little maintenance. It can be pulled by
> small native draft animals which are too weak to pull a
> moldboard plow through the hard, stony soils of Kirman. In
> short, the nail plow is adjusted to local agricultural
> conditions (English 1966:106–107).

All too frequently the "best" American practice of one generation becomes the folklore of the next, and not infrequently we have struggled to teach people in developing countries new practices that we ourselves have subsequently abandoned. Allen, in describing home economics training programs in Syria, run by the Near East Foundation during the interwar years, unwittingly illustrates this kind of hollow "success."

> *Here and there, signs of definite progress began to appear in their [the villagers'] homes. . . . Several of the mothers were gradually learning to nurse their babies on schedule instead of thrusting their breasts into the mouths of the infants at the first whimper (Allen 1943:58; my emphasis).*

Uneducated peoples may be wrong on technical matters; they often are. But until we are sure they are wrong on a particular point, it is unwise and morally wrong to try to "improve" them. It is wrong to assume that a method, because it is modern, scientific, and Western, is better than a traditional one. Read reports to us the folk wisdom of an old woman in a village in Central Africa:

> *You Europeans think you have everything to teach us. You tell us we eat the wrong food, treat our babies the wrong way, give our sick people the wrong medicine; you are always telling us we are wrong. Yet, if we had always done the wrong things, we should all be dead. And you see we are not (Read 1955:7).*

The ethic, then, of helping people change includes knowing what a sociocultural system is and how such systems change. It is not enough to be a competent technician, morally fortified with the unquestioned assumptions of goodness of one's profession. On must be a *responsibly* competent technician, aware that any technical improvement has social and economic consequences that may or may not be deleterious. The responsible technician is the one who is able to help adapt scientific technology and methods to the ecological, social, and economic environment of the developing country, but who does not think that good consists in leading others to do things as he does. Eugene Staley gives a good example of what this means. Hybrid corn, he points out, was developed in the American Midwest. The strains developed there did not do well in Mexico, however, and it took years, using the Midwest methods, to develop the many strains necessary for the different climatic areas of Mexico. Yet the fundamental invention of corn-hybridization, and the principles and methods learned in developing new strains, applied just as much to Mexico as to Iowa (Staley 1954:229–230).

The ethic of helping people change involves caution and restraint in missionary zeal. It means that developmental personnel should be careful not to plan for people, but to work with them in searching for realistic answers to their problems. Particularly as the role of the technical specialist becomes that of the supportive consultant rather than that of teacher, he must learn to restrain his urge to plunge in to "get the job done," and must learn to recognize that his professional gratification must now come from seeing his local counterparts achieve success. More than ever, he must submerge his professional personality and do his utmost to see that the local specialists with whom he cooperates receive the credit for work well done. His success will be measured by the degree to which they are permitted greater and greater authority and responsibility in making decisions and executing projects.

## Ethics and the anthropologist

The anthropologist who works in a technical assistance capacity is faced with many of the same ethical problems encountered by professional specialists, including that of the morality of intervening—even upon invitation—in other societies. Until the mid-1960s most American anthropologists accepted without question the need for an American foreign aid program; as political liberals, they were among the most vociferous in opposing threats by Congress to cut appropriations. The great majority of anthropologists felt it was legitimate use of their skills to study processes of culture change in action programs and to help introduce better health, agricultural, and educational technologies in developing countries. The Society for Applied Anthropology, first in 1949 and subsequently in 1963, formulated codes of professional ethics to guide anthropologists who engaged in these and other applied anthropology programs, and these codes were felt to be sufficient to cover all contingencies. In short, ethical problems seemed fairly obvious, and the answers to them simple. Today the picture has changed drastically; largely as a consequence of Project Camelot (see following) and of the Vietnam War, anthropologists are concerned with the question of professional ethics as never before. In order to understand the present situation, an historical overview of the development of a sense of ethics will be helpful.

Until little more than a generation ago anthropologists were guided by a simple, unwritten ethical code: "An anthropologist is a gentleman (lady)." We saw ourselves in much the same position as the physician, the lawyer, or the minister: privy to a great deal of confidential information that if carelessly handled, might cause embarrassment and hardship to our informants. We took it for granted that we would no more violate the confidence of our informants than of our friends, and we did our best to make sure that none of our actions, including publication, would adversely affect the people we studied. This simple code worked well; the tribal and peasant peoples then studied by anthropologists were so far removed from the middle and upper classes of their countries that it seemed highly unlikely that harm could come to them because of what an anthropologist might do. Without question, and in the interests of scientific accuracy, we gave the names of, and pertinent facts about, our principal informants, and we went to great effort to be thorough in recording geographical information.

British social anthropologists working on problems of colonial administration in Africa and other parts of the British Empire shared the ethical views of American colleagues. During the interwar years they, like their countrymen, accepted the need for a long period of tutelage under more-advanced European rule before native peoples would be able to achieve self-rule. They believed in a just, humane, and informed colonial rule, and they knew that anthropology could provide information on native political, social, economic, and belief systems that would further such a goal. Anthropologists felt that in helping to improve local tribal administrations, and in interpreting the cultures of native peoples to their governments, they were aiding the subject peoples. It was only in the late 1930s that British anthropologists began to learn from African students studying in England that their professional activities were viewed by Africans as helping to prolong a colonial system that had to come to an early end. This new ethical insight led many British anthropologists to abandon the kind of applied anthropology they had earlier done.

American anthropologists were largely spared the ethical dilemma that faced their British colleagues in colonial areas. Only in the administration of the Trust Territories of Micronesia following World War II did American anthropologists play comparable roles. Here most of those who participated felt that the goals of administration were humane, that anthropologists were given great freedom in their work and great freedom to publish, and that the lot of the Micronesians was improved as a consequence of their research.

Meanwhile, World War II had brought American anthropologists into government service in professional capacities on an unprecedented scale. In their desire to defeat Nazi Germany and Japan, Americans were united in a war effort as never since. To use one's skills in any way that would hasten victory was the highest form of patriotism. Anthropologists fully shared this view; probably half of them (possibly more) engaged in war-related work. Many worked in the Foreign Morale Analysis Division in the Office of War Information, in the field known as Psychological Warfare. Others served in Intelligence, and still others used their specialized geographical knowledge to aid in drawing up plans for invasion and other military operations. Many carried out their war roles in uniform, and all worked closely with the military establishment. No criticism of anthropological war-related activities was voiced; quite the contrary, anthropologists were pleased that their skills had military value.

In this way a pattern of government employment and consultation, including Department of Defense ties, was established. With the end of the war and the beginning of major American technical assistance programs, large numbers of anthropologists happily accepted long- and short-term assignments in which they hoped to make more effective the delivery of technical assistance. So, by 1960, and especially because of the Agency for International Development, the government was the single largest employer of applied anthropologists, and by far the largest user of anthropologists for a wide variety of consultantships. Although a majority of anthropologists preferred a good university job as a permanent berth, government service either as a short-term consultant or as a direct-hire employee was honorable and aboveboard.

In the early 1960s the situation began to change, so slowly and subtly that most anthropologists did not at the time appreciate what was happening. The Department of Defense, which had long supported social science research in a variety of ways, stepped up its level of support, and anthropologists and other social scientists, and their universities as well, were drawn into contractual arrangements with government for increasingly dubious purposes. Matters came to a head with Project Camelot, instigated in 1964 by the Army's Special Operations Research Office. Through the mechanism of what was apparently a sham university contract, American social scientists (including an anthropologist) were sent to Chile in early 1965 to enlist the cooperation of Chilean social scientists in survey research to study the general question of identifying criteria for predicting possible revolution, and to find means of dealing with such situations. Chile seems to have been selected, not because of the imminence of revolution, but rather because it seemed a logical scientific site for such a project.

Although Project Camelot was in no sense a secret in Chile, Department of Defense support apparently was concealed. When the U.S. Army's interest in the project was revealed in May, 1965, "Chile was the scene of wild newspaper tales of spying and academic outrage at being recruited for spying missions" (Horowitz 1967:11). The project was canceled, American personnel were called home, and the academic community has never been the same since. Insofar as anthropology was involved, the import was clear: Research involving at least one American anthropologist was being used as a cover for military purposes.

Anthropologists immediately recognized that, quite apart from the basic morality of the project, Camelot threatened the entire profession, since henceforth *any* anthropologist who worked abroad, however legitimate his support, might be suspected of similar, covert activities. In the aftermath of Camelot, the American Anthropological Association, which had never had a formal code of ethics, commissioned a distinguished senior anthropologist and former Association President, Professor Ralph L. Beals, to head a committee to consider the whole broad problem of ethics in research and professional practice. The committee report led to the "Statement on Problems of Anthropological Research and Ethics," which was overwhelmingly approved by Association Fellows in 1967 (American Anthropological Association Fellows 1967). While not a formal code of ethics, since it specifies no sanctions, the statement goes far beyond previous Association positions in spelling out the conditions under which American anthropologists should engage in government work. It particularly stresses the ethical imperative of a full and frank disclosure of all research support, and the right of anthropologists to publish their research findings without government censorship. Except in times of war formally declared by Congress, says the Statement, anthropologists should undertake no work not related to their usual teaching, research, and public service functions, nor should they engage in clandestine activities.

During these same years there was growing frustration among anthropologists and students, as well as among large segments of American society, because of the escalation of the war in Vietnam and their inability to affect the course of events. Some anthropologists continued to serve as government consultants during this time, including service on panels and committees with Department of Defense and military personnel. The extent to which anthropologists have been involved in southeast Asian affairs as a consequence of the Vietnam War, and whether any have violated the letter or spirit of the Statement on Problems of Anthropological Research and Ethics, is unclear. Charges and countercharges, accusations and denials, have shaken the American Anthropological Association since 1970, but we do not know all of the facts, and perhaps we never will. In judging colleagues it is well to remember that during the early years of the war there were still a great many sincere Americans who believed that in aiding Vietnam we were fighting communism, so that war-related professional activities, following the World War II model, were honorable and ethically defensible.

The important point is that out of the crucible of the 1960s there has developed among anthropologists a far keener sense of the complexity of ethical problems in the practice of their profession. The old-fashioned informal codes of protecting informants and patriotically and unquestioningly serving government, adequate in the past, are no longer sufficient. As a consequence of Camelot and Vietnam, and of the new voices of the peoples anthropologists traditionally have studied, it is clear that anthropologists are developing broadened understandings of the ethical dimensions of their profession. The ethical problem increasingly is seen as a dynamic and not an absolute phenomenon; what was appropriate behavior yesterday may not be adequate today. In spite of absence of enforceable sanctions in codes of ethics, the professional behavior of the vast majority of anthropologists has been above censure. I am confident that future anthropologists will adhere to these same high standards.

# works
# cited

**Abou-Zeid,** A.M.
  1963  "Migrant labour and social structure in Kharga Oasis."
       In J. Pitt-Rivers, ed., *Mediterranean countrymen: essays
       in the social anthropology of the Mediterranean*, pp. 41–
       53. The Hague: Mouton & Co.
**Aceves,** Joseph
  1971  *Social Change in a Spanish village*. Cambridge, Mass.:
       Schenkman.
**Adamic,** Luis
  1934  *The native's return*. New York: Harper & Row.
**Adams,** Harold S., George M. Foster and Paul S. Taylor
  1955  *Report on community development programs in India,
       Pakistan and the Philippines*. Washington, D.C.: Interna-
       tional Cooperation Administration.
**Adams,** John B.
  1957  "Culture and conflict in an Egyptian village," *American
       Anthropologist* 59:225–235.
**Adams,** Richard N.
  1951  "Personnel in culture change: a test of a hypothesis,"
       *Social Forces* 30:185–189.
  1953  "Notes on the application of anthropology," *Human Or-
       ganization* 12(2):10–14.
  1955  "A nutritional research program in Guatemala." In B. D.
       Paul, ed., *Health, culture and community*, pp. 435–458.
       New York: Russell Sage Foundation.
**Allen,** Harold B.
  1943  *Come over into Macedonia*. New Brunswick, N.J.: Rutgers
       University Press.

**Altamira,** Rafael
    1949    *A history of Spain from the beginnings to the present day.* Translated by Muna Lee. New York: Van Nostrand Reinhold.

**American** Anthropological Association Fellows
    1967    "Statement on problems of anthropological research and ethics," *American Anthropologist* 69:381–382.

**Ammar,** Hamed
    1954    *Growing up in an Egyptian village. Silwa, Province of Aswan.* London: Routledge & Kegan Paul.

**Apodaca,** Anacleto
    1952    "Corn and custom: the introduction of hybrid corn to Spanish American farmers in New Mexico." In E. H. Spicer, ed., *Human problems in technological change,* pp. 35-39. New York: Russell Sage Foundation.

**Bailey,** Flora L.
    1948    "Suggested techniques for inducing Navaho women to accept hospitalization during childbirth and for implementing health education," *American Journal of Public Health* 38:1418–1423.

**Banfield,** Edward C.
    1958    *The moral basis of a backward society.* New York: Free Press.

**Barker,** Anthony
    1959    *The man next to me: an adventure in African medical practice.* New York: Harper & Row.

**Barnett,** H. G.
    1942    "Applied anthropology in 1860," *Applied Anthropology (now Human Organization)* 1(3):19–32.
    1953    *Innovation: the basis of cultural change.* New York: McGraw-Hill.
    1956    *Anthropology in administration.* New York: Harper & Row.

**Batten,** T. R.
    1957    *Communities and their development: an introductory study with special reference to the tropics.* London: Oxford University Press.

**Beers,** Howard W.
    1950    "Survival capacity of extension work in Greek villages," *Rural Sociology* 15:274–282.

**Benedict,** Ruth
    1940    *Race: science and politics.* New York: Modern Age.

**Berreman,** Gerald D.
   1963     *Hindus of the Himalayas.* Berkeley: University of California
            Press.
**Biesanz,** John and Mavis Biesanz
   1944     *Costa Rican life.* New York: Columbia University Press.
**Bourdieu,** Pierre
   1963     "The attitude of the Algerian peasant toward time."
            In J. Pitt-Rivers, ed., *Mediterranean countrymen: essays
            in the social anthropology of the Mediterranean,* pp. 55–
            72. The Hague: Mouton & Co.
**Brokensha,** David and Peter Hodge
   1969     *Community development: an interpretation.* San Francisco:
            Chandler.
**Brown,** G. Gordon
   1957     "Some problems of culture contact with illustrations from
            East Africa and Samoa," *Human Organization* 16(3):11–14.
**Browning,** Harley L. and Waltraut Feindt
   1971     "The social and economic context of migration to Monter-
            rey, Mexico." In F. F. Rabinovitz and F. M. Trueblood,
            eds., *Latin American urban research,* 1:45–70. Beverly
            Hills, Calif.: Sage Publications.
**Buitrón,** Aníbal
   1959     "La investigación y el mejoramiento de las condiciones de
            vida," *Boletín Informativo* nos. 26–27:6–9. Pátzcuaro, Mi-
            choacán, Mexico: CREFAL.
   1960     "Problemas económico-sociales de la educación en la
            América Latina," *América Indígena* 20:167–172.
**Butterworth,** Douglas S.
   1962     "A study of the urbanization process among the Mixtec
            migrants from Tilaltongo in Mexico City," *América In-
            dígena* 22:257–274.
**Caldwell,** John C.
   1969     *African rural-urban migration: the movement to Ghana's
            towns.* New York: Columbia University Press.
**Caro Baroja,** Julio
   1963     "The city and the country: reflexions on some ancient
            commonplaces." In J. Pitt-Rivers, ed., *Mediterranean coun-
            trymen: essays in the social anthropology of the Mediter-
            ranean,* pp. 27–40.. The Hague: Mouton & Co.
**Carstairs,** G. Morris
   1958     *The twice-born: a study of a community of high-caste
            Hindus.* Bloomington: University of Indiana Press.

**Cassel,** John
 1955   "A comprehensive health program among South African
        Zulus." In B. D. Paul, ed., *Health, culture and community,*
        pp. 15–41. New York: Russell Sage Foundation.
**Chapman,** Charlotte Gower
 1971   *Milocca: a Sicilian village.* Cambridge, Mass.: Schenkman.
**Clark,** Grahame
 1947   *Archaeology and society.* 2nd ed., rev. London: Methuen.
**Clark,** Margaret
 1959   *Health in the Mexican-American culture.* Berkeley: Uni-
        versity of California Press.
**Cobarruvias** [Covarrubias]**Orozco,** Sebastián de
 1611   *Tesoro de la lengua castellana, o Española.* Madrid. (Re-
        printed in Barcelona, 1943, by S. A. Horta, I. E.)
**Coleman,** James S.
 1958   *Nigeria: background to nationalism.* Berkeley: University
        of California Press.
**Collins,** Gretchen E.
 1955   "Do we really advise the patient?" *The Journal of the
        Florida Medical Association* 42:111–115.
**Colson,** Elizabeth
 1958   *Marriage and the family among the Plateau Tonga of
        Northern Rhodesia.* Manchester: University of Manchester
        Press.
**Cousins,** Norman
 1961   "Confrontation," *Saturday Review* 44(12):30–32.
**Crane,** Robert I.
 1955   "Urbanism in India," *The American Journal of Sociology*
        60:463–470.
**Croizier,** Ralph C.
 1968   *Traditional medicine in modern China: nationalism, and
        the tensions of cultural change.* Cambridge, Mass.: Harvard
        University Press.
**Cussler,** Margaret and Mary L. De Give
 1952   *'Twixt the cup and the lip. Psychological and socio-
        cultural factors affecting food habits.* New York: Twayne.
**Davis,** Stanley M.
 1968   "Management's effects on worker organizations in a
        developing country," *Human Organization* 27:21–29.
**Díaz-Guerrero,** Rogelio
 1967   "The active and passive syndromes," *Interamerican Journal
        of Psychology* 1:263–272.

**Dube,** S. C.
  1955   *Indian village.* Ithaca, N.Y.: Cornell University Press.
  1956   "Cultural factors in rural community development." *The Journal of Asian Studies* 16:19–30.
  1957   "Some problems of communication in rural community development," *Economic Development and Cultural Change* 5:129–146.
  1958   *India's changing villages: human factors in community development.* London: Routledge & Kegan Paul.

**Elkin,** A. P.
  1936–37 "The reaction of primitive races to the white man's culture." *The Hibbert Journal* 35:537–545.

**Embree,** John F.
  1946   "Military government in Saipan and Tinian," *Applied Anthropology* (now *Human Organization*) 5(1):1–39.

**English,** Paul Ward
  1966   *City and village in Iran: settlement and economy in the Kirman basin.* Madison: University of Wisconsin Press.

**Erasmus,** Charles J.
  1952   "Agricultural changes in Haiti: patterns of resistance and acceptance," *Human Organization* 11(4):20–26.
  1954   "An anthropologist views technical assistance," *The Scientific Monthly* 78:147–158.
  1955   *Reciprocal labor: a study of its occurrence and disappearance among farming peoples in Latin America.* Unpublished Ph.D. dissertation, University of California, Berkeley.

**Fallers,** L. A.
  1961   "Are African cultivators to be called 'peasants'?" *Current Anthropology* 2:108–110.

**Fei,** Hsiao-Tung and Chih-I-Chang
  1945   *Earthbound China: a study of rural economy in Yunnan* Chicago: University of Chicago Press.

**Firth,** Raymond
  1956   *Elements of social organization.* 2nd ed. London: Watts.

**Flinn,** William L.
  1971   "Rural and intra-urban migration in Colombia: two case studies in Bogotá." In F. F. Rabinovitz and F. M. Trueblood, eds., *Latin American urban research* 1:83–93. Beverly Hills, Calif.: Sage Publications.

**Forde,** Daryll
  1953    "Applied anthropology in government: British Africa."
          In A. L. Kroeber, Chairman, *Anthropology today,* pp. 841–
          865. Chicago: University of Chicago Press.
**Foster,** George M.
  1948    "Some implications of modern Mexican mold-made pot-
          tery," *Southwestern Journal of Anthropology* 4:356–470.
  1952    "Relationships between theoretical and applied anthro-
          pology: a public health program analysis," *Human Or-
          ganization* 11(3):5–16.
  1965    "Peasant society and the image of limited good," *Amer-
          ican Anthropologist* 67:293–315.
  1967a   *Tzintzuntzan: Mexican peasants in a changing world.* Bos-
          ton: Little, Brown.
  1967b   "The Institute of Social Anthropology of the Smithsonian
          Institution 1943–1952," *Anuario Indigenista* 27:173–192.
**Freedman,** Maurice
  1956    "Health education: how it strikes an anthropologist,"
          *Health Education Journal* 14:18–24.
**Friedl,** Ernestine
  1958    "Hospital care in provincial Greece," *Human Organization*
          16(4):24–27.
  1959    "The role of kinship in the transmission of national culture
          to rural villages in mainland Greece," *American Anthro-
          pologist* 61:30–38.
**Friedmann,** F. G.
  1958    "The world of 'La Miseria'," *Community Development
          Review* no. 10:16–28. Washington, D.C.: International
          Cooperation Administration. (Reprinted from *Partisan Re-
          view,* 1953.)
**Gallop,** Rodney
  1936    *Portugal: a book of folk-ways.* London: Cambridge Uni-
          versity Press.
**García, Manzanedo,** Héctor and Isabel Kelly
  1955    *Comentarios al proyecto de la campaña para la erradica-
          ción del paludismo en México.* Mimeographed. México,
          D. F.: Dirección de Estudios Experimentales en Salubridad
          Pública, Instituto de Asuntos Interamericanos.
**Geertz,** Clifford
  1962    "The rotating credit association: a 'middle rung' in de-
          velopment," *Economic Development and Cultural Change*
          10:241–263.

**Germani,** Gino
　1961　"Inquiry into the social effects of urbanization in a work-ing-class sector of greater Buenos Aires." In P. M. Hauser, ed., *Urbanization in Latin America,* pp. 206–233. New York: International Documents Service.

**Gibb,** H. A. R. and Harold Bowen
　1950　*Islamic society and the west: a study of the impact of Western civilization on Moslem culture in the Near East,* vol. 1 pt. 1. London: Oxford University Press.

**Gittler,** Joseph B.
　1949　"Man and his prejudices," *The Scientific Monthly* 69:43–47.

**Gladwin,** Thomas
　1956　"Anthropology and administration in the Trust Territory of the Pacific Islands." In *Some uses of anthropology: theoretical and applied,* pp. 58–65. Washington, D.C.: The Anthropological Society of Washington.

**Goswami,** U. L. and S. C. Roy
　1953　"India." In Phillips Ruopp, ed., *Approaches to community development,* pp. 299–317. The Hague, Bandung: W. Van Hoeve.

**Gough,** E. Kathleen
　1952　"Changing kinship usages in the setting of political and economic change among the Nayars of Malabar," *Journal of the Royal Anthropological Institute* 82:71–88.

**Graubard,** Mark
　1943　*Man's food, its rhyme or reason.* New York: Macmillan.

**Hailey,** Lord
　1944　"The rôle of anthropology in colonial development," *Man* 44:10–15.

**Hamamsy,** Laila Shukry
　1957　"The role of women in a changing Navaho society," *American Anthropologist* 59:101–111.

**Hawley,** Florence
　1946　"The role of Pueblo social organization in the dissemina-tion of Catholicism," *American Anthropologist* 48:407–415.

**Hereford,** Philip, ed.
　1935　*The ecclesiastical history of the English people* by the Venerable Bede. Translated by Thomas Stapleton. London: Burns & Oates.

**Hertz,** Will
　1958　"Mother India," *Oakland (California) Tribune* March 9.

**Hesiod**
  1959    *Hesiod*. Translated by Richard Lattimore. Ann Arbor: University of Michigan Press.
**Hill,** Colin DeN.
  1957    "Some notes on experience of community development in southeast Nigeria," *Community Development Review* no. 7:18–23. Washington, D.C.: International Cooperation Administration.
**Hill,** W. W.
  1944    "The Navaho Indians and the ghost dance of 1890," *American Anthropologist* 46:523–527.
**Hodgkin,** Thomas
  1956    *Nationalism in colonial Africa*. London: Frederick Muller.
**Hofer,** Tamás
  1968    "Comparative notes on the professional personality of two disciplines," *Current Anthropology* 9:311–315.
**Hogbin,** H. Ian
  1958    *Social change*. London: Watts.
**Hollnsteiner,** Mary R.
  1963    *The dynamics of power in a Philippine municipality*. Diliman, Quezon City: University of the Philippines, Community Development Research Council.
**Horowitz,** Irving Louis
  1967    "The rise and fall of Project Camelot." In I. L. Horowitz, ed., *The rise and fall of Project Camelot: studies in the relationships between social science and practical politics*. Cambridge, Mass.: MIT Press.
**Hoyt,** Elizabeth E.
  1956    "The impact of a money economy on consumption patterns." In B. F. Hoselitz, ed., *Agrarian societies in transition*, 305:12–22. The Annals of the American Academy of Political and Social Science.
**Huard,** Pierre and Ming Wong
  1968    *Chinese medicine*. Translated from the French by Bernard Fielding. New York: McGraw-Hill.
**Hunter,** John M.
  1959    "Reflections on the administrative aspects of a technical assistance project," *Economic Development and Cultural Change* 7:445–451.
**Jackson,** I. C.
  1956    *Advance in Africa: a study of community development in Eastern Nigeria*. London: Oxford University Press.

**Jelliffe,** D. B.
  1957   "Social culture and nutrition: cultural blocks and protein malnutrition in early childhood in rural West Bengal," *Pediatrics* 20:128–138.

**Jelliffe,** Derrick B. and F. John Bennett
  1963   "Different problems in different parts of the world," *Acta Paediatrica* Suppl. 151:13–18.

**Jelliffe,** D. B., F. J. Bennett, C. E. Stroud, Hebe F. Welbourn, M. C. Williams and E. F. Patricia Jelliffe
  1963   "The health of Acholi children," *Tropical and Geographical Medicine* 15:411–421.

**Jelliffe,** D. B., J. Woodburn, F. J. Bennett and E. F. P. Jelliffe
  1962   "The children of Hadza hunters," *The Journal of Pediatrics* 60:907–913.

**Joseph,** Alice
  1942   "Physician and patient: some aspects of interpersonal relations between physicians and patients, with special regard to the relationship between white physicians and Indian patients," *Applied Anthropology* (now *Human Organization*) 1(4):1–6.

**Keesing,** Felix M.
  1941   *The south seas in the modern world.* New York: John Day.

**Kelly,** Isabel
  1953   "Informe preliminar del proyecto de habitación en La Laguna, Ejido de El Cuije, cercano a Torreón, Coahuila." Mimeographed. México, D.F.: Instituto de Asuntos Interamericanos.

  1958   "Cambios en los patrones relacionados con la alimentación," *Boletín del Instituto Internacional Americano de Protección a la Infancia,* 32:205–208. Montevideo.

  1960   *La antropologia, la cultura y la salud pública.* Lima, Peru: Imprenta del SCISP (Ministry of Public Health and Social Welfare).

**Kemper,** Robert Van
  1971   *Migration and adaptation of Tzintzuntzan peasants in Mexico City.* Unpublished Ph.D. dissertation, University of California, Berkeley.

**King,** Clarence
  1958   *Working with people in small communities.* New York: Harper & Row.

**Kluckhohn,** Clyde
  1949   *Mirror for man.* New York: McGraw-Hill.

**Kroeber,** A. L.
  1948    *Anthropology.* New York: Harcourt Brace Jovanovich.
**Lange,** Charles H.
  1953    "The role of economics in Cochiti pueblo culture change,"
          *American Anthropologist* 55:674–694.
**Levine,** Donald N.
  1965    *Wax and gold: tradition and innovation in Ethiopian cul-*
          *ture.* Chicago: University of Chicago Press.
**Lewis,** Oscar
  1951    *Life in a Mexican village: Tepoztlán restudied.* Urbana:
          University of Illinois Press.
**Linton,** Ralph
  1936    *The study of man: an introduction.* London: Appleton.
  1945    *The cultural background of personality.* New York:
          Appleton.
**Little,** Kenneth
  1957    "The role of voluntary associations in West African urban-
          ization," *American Anthropologist* 59:579–596.
**Lowenstein,** Susan Fleiss
  1965    "Urban images of Roman authors," *Comparative Studies*
          *in Society and History* 8:110–123.
**Lowie,** Robert H.
  1959    *Robert H. Lowie, ethnologist: a personal record.* Berkeley:
          University of California Press.
**Lundsgaarde,** Henry P.
  1966    *Social changes in the Southern Gilbert Islands: 1938–1964.*
          Mimeographed. Eugene: University of Oregon, Depart-
          ment of Anthropology.
**McCormack,** William C.
  1957    "Mysore villagers' view of change," *Economic Develop-*
          *ment and Cultural Change* 5:257–262.
**McDermott,** W., K. Deuschle, J. Adair, H. Fulmer, and B. Loughlin
  1960    "Introducing modern medicine in a Navajo community,"
          *Science* 131:197–205; 280–287.
**Macgregor,** Gordon
  1964    *Warriors without weapons: a study of the society and*
          *personality development of the Pine Ridge Sioux.* Chicago:
          University of Chicago Press.
**Maddox,** James G.
  1956    *United Nations technical assistance to Mexico.* Mexico,
          D.F.: American Universities Field Staff.

**Mair,** Lucy P.
  1957    *Studies in applied anthropology.* London School of Eco-
           nomics, Monographs on Social Anthropology, no. 16.
           London: University of London, Athlone Press.
**Mandelbaum,** David G.
  1941    "Culture change among the Nilgiri tribes," *American
           Anthropologist,* 43:19–26.
**Mangin,** William P.
  1959    "The role of regional associations in the adaptation of
           rural population in Peru," *Sociologus* 9:23–35.
  1960    "Mental health and migration to cities: a Peruvian case."
           In Vera Rubin, ed., *Culture, society and health,* 84:911–
           917. Annals of the New York Academy of Sciences.
**Marriott,** McKim
  1952    "Technological change in overdeveloped rural areas,"
           *Economic Development and Cultural Change* no. 4:261–
           272. (This subsequently became vol. 1.)
  1955    "Western medicine in a village of northern India." In
           B. D. Paul, ed., *Health, culture and community,* pp. 239–
           268. New York: Russell Sage Foundation.
**Marwick,** Max
  1956    "The continuance of witchcraft beliefs," *The Listener* 55
           (1413):490–492.
**Marx,** Leo
  1964    *The machine in the garden: technology and the pastoral
           ideal in America.* New York: Oxford University Press.
**Matos Mar,** José
  1961    "Migration and Urbanization. The 'barriadas' of Lima: an
           example of integration into urban life." In P. M. Hauser,
           ed., *Urbanization in Latin America.* New York: Columbia
           University Press, International Documents Service. pp.
           170–190.
**Mayer,** Albert and Associates
  1958    *Pilot project, India. The story of rural development at
           Etawah, Uttar Pradesh.* Berkeley: University of California
           Press.
**Menéndez Pidal,** Ramón
  1950    *The Spaniards in their history.* Translated by Walter Starkie.
           London: Hollis & Carter.
**Meyers,** Nechemia
  1971    "A dying beauty spot." *San Francisco Chronicle* November
           24.

**Mitchell,** P. E. (Sir Philip)
   1930   "The anthropologist and the practical man: a reply and a question," *Africa* 3:217–223.
   1951   "Review of Native Administration in the British territories in Africa, by Lord Hailey," *Journal of African Administration* 3:55–65.

**Mosher,** Arthur T.
   1960   *Interrelationships between agricultural development, social organization and personal attitudes and values.* Ithaca: Comparative Extension Publication no. 12. New York State College of Agriculture at Cornell University.

**Murphy,** Robert F. and Buell Quain
   1955   *The Trumai Indians of Central Brazil.* Monograph no. 24. New York: American Ethnological Society.

**Neisser,** Charlotte S.
   1955   "Community development and mass education in British Nigeria," *Economic Development and Cultural Change* 3:352–365.

**Norbeck,** Edward and Harumi Befu
   1958   "Informal fictive kinship in Japan," *American Anthropologist* 60:102–117.

**Oberg,** Kalervo
   1954   *Culture shock.* Bobbs-Merrill Reprint Series in the Social Sciences, A-329. Indianapolis: Bobbs-Merrill.

**Oberg,** Kalervo, and José Arthur Rios
   1955   "A community improvement project in Brazil." In B. D. Paul, ed., *Health, culture and community,* pp. 349–376. New York: Russell Sage Foundation.

**Opler,** Morris E.
   1963   "The cultural definitions of illness in village India," *Human Organization* 22:32–35.

**Opler,** Morris E., and Rudra Datt Singh
   1952   "Economic, political and social change in a village of north central India," *Human Organization* 11(2):5–12.

**Park,** Robert
   1915   "The city: suggestions for the investigation of human behavior in the urban environment," *The American Journal of Sociology* 20:577–612.

**Philips,** Jane
   1955   "The hookworm campaign in Ceylon." In H. M. Teaf, Jr., and P. G. Franck, eds., *Hands across frontiers: case studies in technical cooperation,* pp. 265–305. Ithaca, N.Y.: Cornell University Press.

**Phillips,** Herbert P.
    1965    *Thai peasant personality: the patterning of interpersonal behavior in the village of Bang Chan.* Berkeley: University of California Press.

**Pierson,** Donald
    1955    "Sickness and its cure in a Brazilian rural community." *Anais do XXXI Congreso Internacional de Americanistas,* pp. 281–291. São Paulo.

**Pineda,** Virginia Gutiérrez de
    1955    "Causas culturales de la mortalidad infantil," *Revista Colombiana de Antropología* 4:13–85.

**Pitt-Rivers,** J. A.
    1954    *The people of the sierra.* London: Weidenfeld and Nicolson.

**Read,** Margaret
    1955    *Education and social change in tropical areas.* Camden, N.J.: T. Nelson.

**Redfield,** Robert
    1930    *Tepoztlan: a Mexican village.* Chicago: University of Chicago Press.
    1947    "The folk society," *The American Journal of Sociology* 52:293–308.

**Reichard,** Gladys A.
    1949    "The Navaho and Christianity," *American Anthropologist* 51:66–71.

**Richardson,** F. L. W., Jr.
    1945    "First principles of rural rehabilitation," *Applied Anthropology (now Human Organization)* 4(3):16–31.

**Rosentiel,** Annette
    1954    "Long-term planning: its importance in the effective administration of social change," *Human Organization* 13(2):5–10.

**Schneider,** David M.
    1955    "Abortion and depopulation on a Pacific island." In B. D. Paul, ed., *Health, culture and community,* pp. 211–235. New York: Russell Sage Foundation.

**Sibley,** Willis E.
  1960–61   "Social structures and planned change: a case study from the Philippines," *Human Organization* 19:209–211.

**Silone,** Ignazio
    1934    *Fontamara.* Translated by Michael Wharf. New York: Harrison Smith & Robert Haas.

**Simmons,** Ozzie G.

1955a  "The criollo outlook in the mestizo culture of coastal Peru," *American Anthropologist* 57:107–117.

1955b  "The clinical team in a Chilean health center." In B. D. Paul, ed., *Health, culture and community,* pp. 325–348. New York: Russell Sage Foundation.

1959  "Drinking patterns and interpersonal performances in a Peruvian mestizo community," *Quarterly Journal of Studies on Alcohol* 20:103–111.

**Smith,** Arthur H.

1899  *Village life in China: a study in sociology.* Old Tappan, N.J.: Revell.

**Solien,** Nancie L. and Nevin S. Scrimshaw

1957  "Public health significance of child feeding practices observed in a Guatemalan village," *The Journal of Tropical Pediatrics* 3:99–104.

**Soustelle,** Georgette

1958  *Tequila: un village nahuatl du Mexique oriental.* Travaux et Mémoires de l'Institut d'Ethnologie, no. 62. Paris: Université de Paris.

**Spector,** Paul, Augusto Torres, Stanley Lichtenstein, Harley O. Preston, Johnette B. Clark and Susan B. Silverman

1971  "Communication media and motivation in the adoption of new practices: an experiment in rural Ecuador," *Human Organization* 30:39–46.

**Spicer,** E. H.

1952  "Reluctant cotton pickers: incentive to work in a Japanese Relocation Center." In E. H. Spicer, ed., *Human problems in technological change,* pp. 41–54. New York: Russell Sage Foundation.

**Staley,** Eugene

1954  *The future of underdeveloped countries: political implications of economic development.* New York: Harper & Row.

**Straus,** Murray A.

1953  "Cultural factors in the functioning of agricultural extension in Ceylon," *Rural Sociology* 18:249–256.

**Tannous,** Afif

1944  "Extension work among the Arab Fellahin," *Applied Anthropology* (now *Human Organization*) 3(3):1–12.

**Taylor,** Paul S.

1954  "Can we export 'the new rural society'?" *Rural Sociology* 19:13–20.

**Thompson,** Wallace
    1922    *The Mexican mind: a study of national psychology.* Boston: Little, Brown.
**Ullah,** Inayat
    1958    "Caste, patti and faction in the life of a Punjab village," *Sociologus* 8:170–186.
**Verga,** Giovanni
    1955    *The house by the medlar tree.* Translated by Eric Mosbacher. Garden City, N.Y.: Doubleday Anchor Books.
**von Moltke,** Willo
    1969    "The evolution of the linear form." In Lloyd Rodwin and Associates, *Planning urban growth and regional development: the experience of the Guayana program of Venezuela,* chap. 6. Cambridge, Mass.: MIT Press.
**Wellin,** Edward
    1955    "Water boiling in a Peruvian town." In B. D. Paul, ed., *Health, culture and community,* pp. 71–103. New York: Russell Sage Foundation.
**Whately,** Richard
    1857    *Bacon's essays with annotations.* New York: C. S. Francis & Co.
**Wilson,** Godfrey
    1940    "Anthropology as a public service," *Africa* 13:43–60.
**Wirth,** Louis
    1938    "Urbanism as a way of life," *The American Journal of Sociology* 44:1–24.
**Wiser,** Charlotte Viall, and William H. Wiser
    1951    *Behind mud walls.* New York: Agricultural Missions, Inc.
**Yang** Hsin-Pao
    1949    "Planning and implementing rural welfare programs," *Human Organization* 9(3):17–21.

# index

Abou-Zeid, A. M., 58
  quoted, 58–59
Aceves, Joseph, quoted, 27n.–28n.,
  33–34, 106
Adamic, Luis, quoted, 34
Adams, H. S., et al., 152
Adams, J. B., 21
  quoted, 22
Adams, R. N., 125, 222–225, 237
  quoted, 117, 223–224
Adeniyi-Jones, Adeniyi, 140, 228
Administrators, 234–241
Afghanistan, 93, 97–98, 158, 186
Africa, 65, 87–88, 111–112, 154–156,
  162, 251, 256
    See also under names of coun-
    tries
  agriculture in, 99–100
  changing work patterns in, 60
  nationalism in, 68–69, 74
  nutrition in, 62–63
  peasants in, 31, 38
  urbanization in, 49, 57–60
  volunteer associations in, 50–51, 54
Agency for International Development, 176, 198, 248, 257
Agricultural extension agents, 5, 9,
  24, 92, 94, 127, 135

Agriculture, 1, 92, 99–100, 146, 153–
  154, 159–160, 166–167, 169–170,
  172, 218, 254–255
  innovation in, 157–158, 162, 166
  peasant, 30–31, 97–100
  among the Tanala, 14
Agriculturists, 180, 188
Aid programs. See Technical assistance programs
Alfaro Carlos, 145
Algeria, 84, 171
Allen, Harold B., 162
  quoted, 154, 164, 254
Altamira, Rafael, 201
American Anthropological Association, 258–259
American Indians, 58, 66, 69, 91, 94,
  101, 108, 119, 133–134, 152, 166–
  167, 190, 201–202, 229
Ammar, Hammad, quoted, 134, 157
Analysis
  ongoing, 220–225
  stages of, 215–232
Anthropologists, 25, 28–31, 44, 77, 87,
  98, 106, 117, 134, 162, 177, 191,
  253
  administrators and, 234–241

Anthropologists, (Continued)
  appropriate research techniques
    used by, 203–215
  bureaucracy and, 242–243
  contemporary, 202
  effective, 203
  ethics and, 255–259
  factual knowledge and, 203–204
  in government service, 257–259
  initial steps taken by, 205–206
  point of view of, 203
  role of, 215, 241; in technical aid
    programs, 233–245
  scientific capital of, 240–245
  stages of analyses made by, 215–232
  as troubleshooters, 239
  at work, 197–232
Anthropology, 95, 235
  contemporary applied, 202–209
  proto-applied, 199–202
  second–class, 242
  three contributions of, 203–205
Apodaca, Anacleto, 92
Arabs, 160
  authority and, 173
  food among, 69
Arcadian Myth, the, 27n.–28n.
Archeology, nationalism and, 70–71
Argentina, 46–47, 53, 149, 228
Art, nationalism and, 70
Asia
        See also under names of coun-
          tries
  changing work patterns in, 60
  peasants in, 31, 34, 39
Attitudes
  changing, 9, 23
  as cultural barriers to change, 82–93
  understanding of, 198
Audiovisual communication, 220–222
Authoritarianism, 232
Authority
  loci of, 77, 118–126
  parental, 119–120
  in planned change, 171–174
  power and, 172
  use of, 173
Autocracy, 250
  defintion of, 93–94

Bacon, Francis, quoted, 83
Bailey, Flora L., 91, 119
Banfield, Edward C., quoted, 33
Barker, Anthony, quoted, 136–137, 139
Barnett, H. G., 123–125, 202, 217–218
  quoted, 124, 243–244
Baroja, Caro, 26
  quoted, 26–27

Barrios, 106, 136
Batten, T. R., quoted, 169–171
Beals, Ralph L., 258
Bede, the Venerable, 199–200
Beers, Howard W., 154
Behavior
        See also Conduct
  appropriate to situation, 23
  changing forms of, 11, 90, 149
  individual, 10, 12, 18–19
  learned, 138, 232
  limited-good, 38
  motivation of, 19
  paternal, 22–24
  patterns of, 12
  peasant, 36, 38, 106
  reciprocal, 107
  standards of, 155
Benedict, Ruth, 152
Berreman, Gerald D., quoted, 135
Birth control, 247
Birth rates, 1, 3, 46
Blair, Hazel I., 100, 119
Body positions, customary, 101–103
Bogotá, Colombia, 48, 89
Bolivia, 99, 149, 250
Bourdieu, Pierre, quoted, 84, 191
Brazil, 85–86, 113, 162
  community development in, 79–80
British Columbia, 201
Brokensha, David, and Peter Hodge,
    quoted, 38
Brown, G. Gordon, 128
  quoted, 108–109, 129
Browning, Harley L., and Waltraut
    Feindt, 49
Buenos Aires, 46–47, 53
Buitrón, Anibal, 98
  quoted, 168
Bureaucracies, 21, 77, 175–196, 198,
    226, 242–243
  as cultures, 178–179
  premises in, 179–180, 197
Bureaucratic barriers, 175–180
Bureaucratic structure, 177, 180
Bureaucrats, 178, 198–199
Burgess, Ernest, 28
Butterworth, Douglas S., 49–51
  quoted, 53

Caldwell, John C., 49
  quoted, 48
California, 119, 136, 209
  University of, 203
Canitrot, Carlos, 228
Carstairs, G. Morris, quoted, 34
Cassel, John, 160, 162
Caste, 77, 126–127

Caste barriers, 127–128
Catholicism, 94
Ceylon
  agricultural extension service in, 127
  antihookworm campaign in, 231–232
  public health program in, 79–80, 137
Change
  barriers to, 77–78, 150, 231; cultural. *See* Cultural barriers; psychological. *See* Psychological barriers; social. *See* Social barriers
  basic conditions for. *See* Stimulants to change
  cultural. *See* Cultural change
  culture and, 76–78; examples of, 78–80
  dynamics of, 76–81, 149, 171
  factors facilitating, 78
  fear of, 65
  forces for, 76
  historical, 14
  induced, 253
  planned, 9; ethics in, 246–259; sociocultural, 11
  promotion of, 77
  stimulants to. *See* Stimulants
  unplanned, 9
  value of 78
Chapman, Charlotte Gower, quoted, 84, 135
Chicago, urban research in, 28
Children
  behavior patterns of, 23–24
  fathers and, 22
Chile, 90, 137, 149, 227, 230, 250
  health education in, 213–214
  and Project Camelot, 257–259
China, 34, 165
  medicine in, 19, 71–73
Christianity, 93–94, 126, 199–202
Churches, 128–129, 199
Cities, 26
      *See also* names of cities
  attractions of, 45–47
  as focal ponts for change, 56–58
  migration to, 3, 44–63, 149; causes of, 46–47
  new, 149
  peasant communities and, 31–33
  route to, 47–49
  settling in, 49–50
  social life in, 49
City life, 3, 26–27, 44–45
  quality of, 52–54
  urban-rural ties and, 54–55

Civilizations
  complexity of, 16
  traditional, 74
Clans, 106
Clark, Grahame, quoted, 70–71
Clark, Margaret, 119–121, 135, 143, 209
  quoted, 136
Class, 227
  as factor in social change, 77
Class barriers, 127–128
Clothing
  modesty and, 90–91
  nationalism and, 70
  prestige value of, 155–156
Cognition
  behavior and, 18–21
  individual, 10
Coleman, James S., quoted, 57, 59
Collins, Gretchen, E., 142–143
Colombia, 48, 85, 87, 89, 133, 250
Colorado River Relocation Center, 113
Colson, Elizabeth, quoted, 110–111
Communication, 77, 131
  faulty, 131
  problems in, 141–146, 191
  study of, in Ecuador, 220–222
  verbal, 142–143
      *See also* Speech
Communism, 248
Communities, 10
  natural, 178
  peasant. *See* Peasant communities
  rural. *See* Rural communities
  social structure of, 40
  traditional, 105
Community developers, 180
Community development
  good, 184
  international, 203
  philosophy and methodology of, 182–183
  program orientation in, 182–184
  technological, 197
Community Development Programme,
  India, 96–97, 114, 128, 231
Community development projects, 160–161, 176
  in Brazil, 79–80
  in India, *See* India
  literacy, adult educaton, and, 167–169
  working in, 199
Community study; example of, 209–211
Competition; as motivation to change, 152, 158–160, 163

Conduct, group, 10. See also Behavior
Conflict, 115–118
  factionalism and, 115–117
Cook Islands, 102
Cooking, 102
  peasant, 96–97
Costa Rica, 133
Countries. See also under names of countries
  developing. See Developing countries
  industrialized, 3, 59, 149
Cousins, Norman, quoted, 7
Covarrubias, Orozco, quoted, 83
Crane, Robert I., quoted, 56
Credit associations, 52
Crozier, Norman, quoted, 72–73
Cuba, 149, 250
Cultural barriers to change, 77, 82–104
  culture structure as, 93–104
  ethnocentrism as, 86–88
  fatalism as, 85–86
  norms of modesty as, 90–91
  pride and dignity as, 88–90
  relative values as, 91–93
  tradition as, 82–84
  values and attitudes in, 82–93
Cultural change, 9
  barriers to. See Cultural barriers to change
  directed, strategy of, 97
  isolation and, 95
  social cost of, 95
Cultural ethnocentrism, 86–88
Cultural forms, cognition and, 18–21
Cultural pattern, total, relation of institution to, 205
Cultural relativism, 87, 253
Cultural shock, 190–196
  causes of, 191–194, 196
  duration of, 195
  immunity to, 192, 196
  symptoms of, 191
  tourists and, 194
Culture traits, incompatibility of, 93–94
Cultures, 5–6
  American, 181
  analysis of, 202
  bureaucracies as, 178–179
  change and. See Culture change
  defined, 11
  donor, 177
  essence of, 86–87
  function of, 22
  learned, 12

  nationalism and, 66, 73–74
  peasant, 39
  professional, 177
  race and, 12
  reaction to technicians and, 198
  religion and, 13
  social interaction and, 21–24
  society and, 10, 13, 23, 232
  stability of, 76
  structure of, 93–101
  value systems and, 18
  world, America's debt to, 16–18
Currie, L., 90
Cussler, Margaret, and Mary L. De Give, 156–157
Customs, 13, 16
  changing, 9
  of strange country, acceptance of, 193–194

Dancing, 70
Davis, Stanley M., 52
DDT, 100, 219
Death control, 247
Death rates, 1, 3, 46
Decision making
  group, 120–121
  individual, 118–121
  right of, 121
Democracy, 27, 231
  autocracy and, 93–94
  in small communities, 122
Demonstrations, communication dangers in 144–146
Developing countries, 3–8, 251
  competition in, 160
  government policy in, 149
  governments of, 249
  migration to cities and, 47
  nationalism and, 66
  social change in, economics of, 80–81
  technical assistance programs for, 181–182
  technicological development in, 9
Development
  national, 3
  sociotechnical, 4
  technological. See Technological development
Deviants, 124–125
Diaz-Guerrero, Rogelio, quoted, 20
Dietary deterioration, 160
  urbanization and, 61–63
Dietary patterns, 187
  See also Food
  nationalism and, 68–70
Dignity, as barrier to change, 88–90

Disciplines
  professions and, 234–235
  theories and, 235
Disease, control of, 3, 46, 79, 100
Dress. *See* Clothing
Dube, S. C., 103–104, 153–155, 162, 171–173
  quoted, 57, 88, 93, 104, 116, 122–123, 128, 146, 156, 159, 161, 174
Duncan, William, 201–202
Dyadic contrast, 51, 105–106
Dynamics of change, 76–81, 149, 171
  small-group, 111–114

Eating habits, 92–93
*Ecclesiastical History of the English People, The,* 200
Economic gain, desire for, 152–155
Economic reality, 164
Economy, the
  monetary, growth of, and family structure, 58–60
  peasant, 35, 38
  subsistence, 38–59
Ecuador, 98, 168, 229
  experiment in, 220–222
Education, 4–5
  adult, 167–169
  health, 143–145, 165, 213–214
  in nutrition, 103–104
Educators, 188
Ego gratification, 188–190, 235–236, 239
Egypt, 21–22, 58, 84, 86, 132, 137, 157, 173
  social work in, 134
Eisenhower, Dwight D., 111
El Salvador, 103, 228–229
Elites, 2, 250
  culture and, 66
Elkin, A. P., 65
Elson, Benjamin and Adele, 132
Embree, John F., quoted, 190
Employment, as cause of migration to cities, 46–47
English, Paul Ward, quoted, 253
Environment, 12, 35
  changing, 21
Epidemics, 230
Erasmus, Charles J., 61, 154, 158, 226, 229–230
  quoted, 135, 141
Eskimos, 86, 100–101, 119
Ethics
  anthropologists and, 255–259
  in planned change, 246–259
  in technical assistance, 247–252
Ethiopia, 69

Ethnocentrism, cultural, 86–88
Ethnographic growth, 64
Europe, 3, 152
  *See also* under names of countries
  nationalism in, 64, 67
  rural communities in, 42
Europeans, 86
Evaluation
  examples of, 225–232
  importance of, 225
Expectations, 105
Exploitation, fear of among peasants, 37

Factions, social change and, 77
Families
  authority within, 118–121
  obligations and expectations within, 105–111
  peasant, 36, 39–40
  relationships in, 77
Family structure, changes in, 58–60
Farland, Merle S., 144
Farms, single-family, 27
Fatalism, as barrier to change, 85–86
Father-child relationship, in Latin America, 54
Fathers, behavior of, 22–24
Fei, Hsaio-Tung, and Chih-I-Chang, 20
  quoted, 19
Fenton, W. N., 167
Firth, Raymond, 165
Flinn, William L., 48
Folklore, 70
Food, 61–63
  *See also* Dietary patterns
  cooking of. *See* Cooking
  health and, 69–70
  role of, 68
  superstitions about, 104
  traditional, 69
Forde, Daryll, quoted, 238
Foster, George M., 51, 102, 105, 226
Freedom, personal, 2
Friedl, Ernestine, 159–160, 211
  quoted, 57, 157, 212–213
Friedman, Maurice, quoted, 33
Friendship, 106, 116
  formalized, 40
  obligations of, 152, 160–162

Gadalla, Fawzy, 86, 132, 137, 164
Gallop, Rodney, quoted, 139
Garcia, Victoria, 137
Geertz, Clifford, quoted, 52
Germani, Gino, quoted, 46–47, 53

Ghana, 49, 168, 170
Gibb, H. A. R., and Harold Bowen, quoted, 173
Gifts, perception of, 136–138
Gilbert Islands, 109–110
Gittler, Joseph B., quoted, 86
Gladwin, Thomas, quoted, 244–245
Godparenthood system, 40
Good, limited, 21
    image of, 35–41
    premise of, 179
    unlimited, premise of, 179
Good Neighbor Policy, 248
Goswami, U. L., and S. C. Roy, quoted, 117–118
Gough, E. Kathleen, quoted, 59–60
Government
    of developing countries, 249
    perception of role of, 134–136
    responsiveness of, to people, 2
    suspicion of, 134–136, 231
    U.S. military, 190
Granbard, Mark, quoted, 172
Gratitude, for technical assistance, 190
Great Britain, and colonialism, 256
Greece, 154, 157, 159, 162–163, 167
    ancient, 26–27
    rural hospital in, 211–213
    upward mobility in, 57
Green Revolution, the, 246–247
Greenland, 86
Gregory, Pope, quoted, 200
Group solidarity, 106–115
Groups, 10
    primitive, 121
    small, dynamics of, 111–114
    social, variation among, 12
Guatamala, 117, 174, 237
    case study of health program in, 222–225
Guevara, Che, quoted, 84
Guilt feelings, 36
Gupta, Harry, 74
Gutierrez de Pineda, Virginia, quoted, 85, 87

Hailey, Lord, quoted, 237
Haiti, 60, 95–96, 141
Hamamsy, Laila S., 58
Hawley, Florence, 94
Health, 131, 247
    public. See Public health
Health centers, 226–227
Health problems, 132–134, 209–214, 235
Health programs. See Public health
Herder, Johann Gottfried, 64

Hereford, Philip, 200–201
Hertz, Will, quoted, 114–115
Hesiod, quoted, 28
Hill, Colin, quoted, 122, 159
Hill, W. W., quoted, 94
Hodgkin, Thomas, quoted, 74
Hofer, Tamás, quoted, 64–65
Hollensteiner, Mary R., quoted, 116
Hopkirk, Mary, 102–103
Horowitz, Irving Louis, quoted, 258
Hospitality, 116
Hospitals, 137–139, 157, 229–230
    fear of, 132–133
    rural, in Greece, 211–213
Hoyt, Elizabeth E., quoted, 156
Hozbin, H. Ian, 156
Huard, Pierre, and Ming Wong, quoted, 72
Huenemann, Ruth, 204
Human relations, in bureaucracies, 179
Humor and wit, national, 71
Hungary, 65
Hunter, John M., 127

Ideas, new, presentation of, 131
Identification, small-group, 106, 111
Illiteracy, 1
Illness, 1
    See also Disease
    social roles of, 209–211
India
    agriculture in, 146, 153, 159, 170, 218
    authority in, 124–125
    city life in, 56–57
    Community Development Programme in, 96–97, 114, 128, 160–163, 231
    factionalism in, 116–117
    innovation in, 173
    kinship change in, 59–60
    leadership in, 122–123
    medicine in, 71–72, 135
    modern, 155–157
    nationalism in, 71–72, 74
    peasants in, 34, 39, 96, 168
    public opinion in, 114–115
Individuals
    change and, 149–150
    decision making by, 118–119
    unusual, authority of, 118, 123–126
Industrialization, 3, 59, 149
Innovation, 147, 150, 176–177
    acceptance of, 173
    middle-class, 170–171
    planned, consequences of, 95–101

Innovation, (Continued)
  receptivity to, authority and, 171–174
  resistance to, 103
  sequence in, 167–169
  successful, 164, 218
  timing and, 169–170
Institute of Inter-American Affairs, 226, 248
Institute of Nutrition of Central America and Panama (INCAP), 222–223
Institutions, 10, 13
  related to total culture pattern, 205
  social, 106
  traditional, roles and, 164–165
Iran, 96–97, 184–185
Israel, 95
Italy, 84
  peasants in, 33–34, 39, 135

Jackson, I. C., 160, 165, 168
  quoted, 151, 159, 180
Japan, 59
Japanese, the, 113–114, 217–218
Jefferson, quoted, 27
Jelliffe, D. B., 156
  quoted, 62–63
Jelliffe, D. B., and F. John Bennett, quoted, 63
Joseph, Alice, quoted, 190

Keesing, Felix M., 125
Kelly, Isabel, 88–89, 165, 206–209, 216, 220, 226
Kemper, Robert Van, 47
  quoted, 53–54
Kim, Y. S., 119
Kin, fictive, 107
King, Clarence, quoted, 134
Kinship organization, 59
  African, 51
Kluckhohn, Clyde, quoted, 23
Knowledge, access to, 1
Kroeber, A. L., quoted, 30

Land, distribution of, 2
Land reform, 249
Landlords, feudal, 118
Lange, Charles H., 58
Language, 141–142
  See also Speech
  difficulties with, 142–144
  grammar and, 19
  as symbol of nationalism, 67–68
Latin America, 133, 157, 204
  See also under names of countries

changing work patterns in, 61
city life in, 52–54
development programs in, 158
fatalism in, 85–86
health program evaluation in, 225–231
migration to cities in, 46–47, 49
nationalism in, 68
peasants in, 31, 34, 39, 98, 111, 149, 158
power and authority in, 172
research study on society and culture of, 241
technical aid to, 248–252
voluntary associations in, 51
Leadership, 182
  dependence on, 123
  patterns of, 121–122
  in peasant communities, 37–38, 183
Learning, 131
  problems in, 146–147
Levine, Donald M., quoted, 69
Lewis, Oscar, 33, 51
  quoted, 29
Life experiences, 12
Lima, Peru, 47–48, 51, 53
Linton, Ralph, quoted, 12, 14, 16–18
Literacy, 167–169
Little, Kenneth, quoted, 50–51
Living standards, 149
Loss of face, 88–89
Lowenstein, Susan, 26
  quoted, 27
Lowie, Robert H., 130
Lundsgaarde, Henry P., quoted, 109–110

Macgregor, Gordon, quoted, 108
Madagasgar, 14
Maddox, James G., quoted, 80
Mair, Lucy P., 154
  quoted, 48, 155
Malaria eradication program, planning of, 218–220
Man
  activities of, analyzed, 10
  ideal, 37–38
  marginal, 123–125
  Western, 26
Mandelbaum, David G., 124
  quoted, 125
Mangin, William P., quoted, 51, 53
Manzanedo, Hector Garcia, 218, 220
Marriage, 87–88
Marriott, McKim, 218
  quoted, 119–120
Martins, Deolinda de Costa, 112
Marwick, Max, quoted, 126

Mayer, Albert, and Associates, quoted, 163
Medical personnel, American, 180
Medical services, 1–2, 91, 174, 190, 247
  poor communication in, 142–144
Medicine, 157
  folk, 226–228
  nationalism and, 71–73
  preventive, 230
Mediterranean countries, 106
      See also under names of countries
  peasants in, 31, 33, 39
Melanesia, 66, 166
Merit, recognition of, 189
Mestizo, the, 31–32
Mexican Americans, 211
Mexico, 105, 142, 146–147
  agriculture in, 166, 254–255
  case history of housing project in, 206–209, 216
  cultural change and, 78–81, 88–89, 100
  dietary deterioration in, 61–62
  food in, 69
  fundamental education project in, 147, 161, 165
  health services in, 78–81, 132–133
  Indians in, 101, 130, 132
  land reform in, 249
  malaria eradication project in, 218–220
  nationalism in, 69–71
  peasants in, 28–29, 32–33, 55, 149
  pottery making in, 102
  progress in, 250
  urbanization in, 46, 49–52, 149
Mexico City, 46–47, 49–52, 54, 227
Meyers, Nechemia, 95
Micronesia, 91
Micronesian Trust Territory, 243–244, 256–257
Middle class, receptivity of, to innovation, 170–171
Middle East, 160, 164
  peasants in, 31
Missionaries, 200–201
Mitchell, Sir Philip, quoted, 235, 237–238
Mobility, upward, 57, 179
Modesty
  female, 91
  norms of, as barrier to cultural change, 90–91
Mosher, Arthur T., quoted, 157
Motivation
  behavior and, 19

intensity of, 153
social change and, 77
as stimulant to change, 151–164
Motor patterns, 101–103, 164–167
Murphy, Robert F., and Buell Quain, quoted, 162

Nasharty, Hala, 84
National Research Council, 203–204
Nationalism, 44, 64–75
  culture and, 66–67
  historical and political, 64–65
  progress and, 75
  successful, 67
  symbols of, 67–75
Nativism, 66
  pagan, 94
Near East, 158
Near East Foundation, 254
Negroes, position of, 127
Neighborhood units, 106
Neisser, Charlotte S., quoted, 135, 160
Nepal, 94, 100
Nervous tensions, 45
New Mexico, 92
New Zealand, 144
Nigeria, 57, 59, 110–111, 122, 140, 159–160, 165, 168, 228
Norbeck, Edward, and Harumi Befu, quoted, 59
Novelty, 82–84
      See also Innovation
  definition of, 83
Nutrition, education in, 103–104. See also Food

Oberg, Kalervo, and José A. Rios, 79–80, 226
  quoted, 191–192
Obligations, 105
  of friendship, 152, 160–162
  mutual, 106–111, 116
Oceania, 66
Opler, Morris E., quoted, 71–72
Opler, Morris E., and Rudra Datt Singh, quoted, 88
Opportunity, 5
  equal, 2
Options open to villagers, 43
Organization of American States, 89
Orientations, cognitive, behavior and, 18–21

Pakistan, 118, 126, 146, 157, 163, 172
Parents, behavior pattern of, 23
Park, Robert, 28
  quoted, 45–46
Peace Corps Volunteers, 7

Peasant communities, 30, 106, 115
  cooperation in, 37
  guilt feelings in, 36
  leadership in, 37–38, 183
  wealth in, 37
Peasant economy, 35, 38
Peasant farmers, 24, 61
Peasant society, 30–35, 167
  cultural change and, 84
  public opinion in, 36
  traditional, 38, 158
Peasant villages, characteristics of, 183
Peasants, 2, 96
  behavior of, 23, 36, 38, 155
  beliefs of, 228–229
  character of, 28–30, 33, 40
  fears of, 40–41
  life of, 30–35
  life strategy of, 38
Peoples
  monotheistic, 93
  primitive, 86
  recipient, 176–177, 189–190, 197–198, 247
Perception, 77, 130
  differential cross-cultural, 131–141
  of purpose, 140–141
Personal relations, overt, 21–22. See also Human relations
Personality, 12–13
  determination of, 20
Personnel
  American medical, 180
  public health, 236, 247
Peru, 34, 48, 51, 53–54, 125–126, 149, 250
  culture of, 66–67
  health program in, 144–145, 162
  nationalism in, 69–71, 73–74
Philippines, the, 116–117, 123, 153
Phillips, Herbert, quoted, 119–120
Phillips, Jane, 231
  quoted, 79, 137
Physicians, female, 91
Pierson, Donald, quoted, 85–86
Pitt-Rivers, J. A., quoted, 161
Planning, 218–220
Play motivation, 162–163
Political groupings, 106
Political structure, authority within, 118, 121–123
Politicians, village, 123
Population growth, migration to cities and, 46
Portugal, 112, 139
Poverty, 1–2, 45, 155
Power, authority and, 172

Premises, 20, 35–36, 39
  in bureaucracies, 179–180, 197
  out-of-date, 21
  sound, 20
  traditional, 20–21
Prestige, 158
  desire for, 152, 155–158
Pride, as barrier to change, 88–90
Primitive peoples, 86
Production, 1
Professions, disciplines and, 234–245
Program orientation, 181
  in community development, 182–184
Progress, discouragement of, 39
Project Camelot, 255, 257–259
Protestantism, 152
Psychological barriers, 77–78, 130–147
  communication problems as, 141–146
  differential cross-cultural perception as, 131–141
  learning problems as, 146–147
Psychology, 10, 199
Public health, 4, 9, 15, 157, 160, 169, 171–172, 174, 175, 187, 216
  in Ceylon, 79–80
  evaluating programs of, 225–232
  in Guatamala, 222–225
  in Iran, 184–185
  in Mexico, 78–80
  in San Jose, California, 209–211
  international, 203
Public health personnel, 236, 247
Public opinion, 114–115
  in peasant society, 36
Purpose, differing perception of, 140–141

Quain, Buell, 162

Radio, 49, 220–222
Read, Margaret, 65
  quoted, 254
Recipient peoples, 176–177, 189–190, 197–198, 247
Reciprocity, social, 105
Redfield, Robert, 28–29
Reichard, Gladys A., 94
Relationships
  family, 77
  father-child, 54
  interpersonal, 106, 226; anthropologist's knowledge of, 205
  personalistic, 40
  social, in an innovating organization, 213–214

Religion, 126, 128–129
  culture and, 13, 93–94
  as motivation to change, 163–164
Religious beliefs, 226
  fatalism and, 86
  motivation to change and, 152
Rendón, Vicente And Natividad, 106
Renteln, Henry, 96–97
Research, 177, 216–218, 239, 241, 258
Revolutions, 2
Rhodesia, Northern. *See* Zambia
Richardson, F. L. W., Jr., 167
Risueño, General Alfredo A. R.,
  quoted, 73–74
Roads, migration and, 48
Rockefeller, Foundation, 79, 231–232
Rockefeller, John D., 137
Role conflict, 128
Role perception, differential, 138–
  140
Role performance, 189
Roles, 5, 11, 13, 22
  behavior and, 23–24
  traditional, 164–165
Rome, ancient, 27
Roosevelt, Franklin D., 248
Rosa, Franz, 131–132
Rosentiel, Annette, 100
Rural communities, 25–47
      *See also* Peasant communities
  contemporary, 42–75
  nutrition in, 61–63
  social change and, 78–81
  voluntary associations and, 50–52
Rural cooperative work patterns, de-
  struction of, 60–61
Rural life, 26–28

Salcedo, Santiago Renjifo, 133
Samoa, 108–109, 128–129
San Jose, California, public health
  study in, 119, 209–211
San Salvador, 112
Sanitation, environmental, 1, 15, 79,
  103, 163, 176
Santiago, Chile, 213
Santos, Marina Beatriz Cruz, 6
Schneider, David M., 91
Science
  behavioral, 202, 216, 225; and tech-
    nical aid, 233–245
  social, terminology of, 237
  Western, 86
Scientific capital, 240–245
Scientists
  behavioral, 197–199, 203, 215, 225–
    226, 233; administrators and,
    234–245

social, 2, 8–9, 29, 52, 242, 257–258
Seghal, G. S., 135
Semi-villagers, 43
Shahin, Zeinab, quoted, 86
Sibley, quoted, 123
Silone, Ignazio, quoted, 34
Simmons, Ozzie G., 214, 226
  quoted, 34, 66–67, 71
Slums, 45
Smith, Arthur H., quoted, 34
Smithsonian Institution, 225–226
  Institute of Social Anthropology,
    241
Social barriers to change, 77, 105–
  129
  conflict as, 115–118
  group solidarity as, 106–115
  loci of authority as, 118–126
Social change, implications of ur-
  banization for, 55–63
Social forms, 164–165
Social groups, man's activities in, 10
Social life
  in cities, 49
  social interaction and, 21
Social science, 237
Social scientists, 2, 8–9, 29, 52, 242,
  257–258
Social structure, 10, 77
  barriers to. *See* Social barriers
  changing, 9
  characteristics of, 126–129
  understanding of, 40
Socialization, 23
Societies, 10
  industrialized, 82–83
  nonindustrial, 85
  peasant. *See* Peasant societies
  primitive, 30, 253
  religion and, 13
  stratified, 127–128
  tension in, 76
  traditional, 1–3
Society, 9–10
  advancement of, 16
  American, 181
  analysis of, 202
  basic configuration of, 128–129
  culture and, 10, 13, 23, 232
  definition of, 11
  nationalism and, 66
  rapidly changing, 23
Society for Applied Anthropology,
  255
Sociocultural forms, learned, 12–13
Sociocultural systems, 77
  changing of, 11–12, 16–18
  characteristics of, 12–20

Sociocultural systems, (Continued)
  defined, 13
  understanding of, 13–14
Sociological premises, 20
Sociologists, 47, 226
  modern, 202
  urban, 28–29
Sociology, 199
  urban, 45, 52
Solien, Nancie L., and Nevin S. Scrimshaw, 174
Soraya, Meydy, quoted, 185
Soustelle, Georgette, 130
Spain, 84, 106, 201
  peasants in, 33–34, 161
Spector, Paul, et al., quoted, 222
Speech, 19
    See also Language
Speech patterns, 101
Spicer, E. H., 113
  quoted, 114
Staley, Eugene, 254–255
Status, 5, 10–11, 13, 22, 128, 227
  peasant, 36, 155
Stimulants to change, 148–174
  motivated, 151–164
Straus, Murray A., 127
Subrahmanyam, Y., 97–98
Suksamiti, Yutthana, 84
Superiority, cultural, 86–87
Superstitions, 103–104
Suspicion
  of gifts, 136–138
  of government, 134–136, 231
Symbols, 141, 144
Syria, 254
Systems
  closed, 38
  peasants and, 35, 38

Taiwan, 89
Tanala, the, 14–15
Tanda, Mexican, 52
Tanganyika, 108–109
Tannous, Afif, 158, 160, 164
  quoted, 159
Tanzania, 62–63
Taylor, Paul S., 95–96
Teamwork in technical aid, 233–245
Technical aid, behavioral science and, 233–245
Technical assistance programs, 8, 21, 28, 43, 98, 136, 175, 180, 257
  American, 179–190
  anthropological model of, 176–177
  donor-culture model of, 177–178
  ethics and, 247–252

evaluation of, 225–232
  ongoing analysis of, 220–225
  overdesign in, 185–187
  planning for, 218–220
  primitive model of, 175–177
  recipients of, 176–177, 189–190, 197–198, 247
  research and, 177, 216–218
  stages of analysis for, 215–232
Technical specialists, 10–11, 21, 75, 77, 91, 95, 120, 135, 177, 198, 251
  adjustments necessary for, 191
  American, 7–9, 25, 28, 103, 163, 180–190
  anthropologists and, 199
  culture shock and, 191–196
  ego-gratification needs of, 188–190
  ethical responsibilities of, 252–255
  problem-oriented, 181–182, 187
  professional pride of, 185, 187–188, 195
  program-oriented, 181–182, 184, 187
  rural community and, 25, 28, 57, 94
  sense of humor needed by, 193
  training of, 180–182, 184
Technological development, 2–8, 188, 215
  cultural context of, 9–24
Technology, Western, 86
Television, 49
Tepoztlan, 28–29
Thailand, 84
Thompson, Wallace, 166
"Three Systems, The," 10
Tiommy, Banción, 113
Tradition, 128
  as barrier to change, 82–84
Transportation, modern, 48
Travel, modern, ease of, 48–49
Trejos, Chacón, quoted, 69–70
Tribal associations, African, 54–55
Truk, 244
Tucson, 229
Tzintzuntzan, 32–33, 46, 53, 55, 78–79, 105, 107, 144, 146–147 150–152

Uganda, 63, 88
Ullah, Inayat, quoted, 172
Underemployment, 3
Unemployment, 3
United Nations, 8, 176
U.N. Educational, Scientific, and Cultural Organization (UNESCO), 147, 161, 165, 176

U.N. Food and Agricultural Organization (FAO), 176
United States, 3, 8, 68–69, 162
  community development and, 182–184
  nationalism in, 74
  Negroes in, 127
  prestige in, 156–157
  shortcomings of, 194
U.S. Army, Special Operations Research Office, 257
U.S. Department of Defense, 257–259
U.S. Department of the Interior, 243
U.S. Navy, 244
U.S. Office of War Information, Foreign Morale Analysis Division, 257
U.S. Public Health Service, 226
Upward mobility, 57, 179
Urban life. See City life
Urban-rural ties, 54–55
Urbanism, definition of, 44
Urbanization, 3, 25, 44–63, 149
  implications of, for social-change, 55–63

Vaccination, 171, 230
Value judgments, 111
Value systems, cultures and, 18
Values, 111
  changing, 18
  cultural change and, 82–93
  peasant, 30
  relative, 91–93
  social and cultural, 164–165
Venezuela, 186–187
Verga, Giovanni, quoted, 84
Vested interests, 117–118
Vietnam War, 255, 258–259
Village autonomy, 129
Villagers
  cooperation among, 60–61, 161
  Indian, 117–118

and literacy, 167–168
  modern, 42–43, 158
Villages. See Rural communities; Peasant communities
Vitamin deficiency, 62
Vountary associations, role of, 50–52, 54
von Moltke, Willo, quoted, 187

Wage labor, 58
Wages, city, 46
Water supply, 15
Wealth, in peasant communities, 37
Wellin, Edward, 125–126, 162
West, the, 3, 5, 66, 86, 119
Whateley, Richard, quoted, 84
Wilson, Godfrey, quoted, 242
Wirth, Louis, 28
  quoted, 44–45
Wiser, Charlotte V., and William H., quoted, 32, 36–37, 39–41
Woleai, 217
Women
  modesty of, 91
  role of, in India, 128
Work
  menial, 89
  professional, 235
Work patterns, rural, 60–61
World, the
  contemporary, rural communities in, 42–75
  traditional, 25–41
World Health Organization (WHO), 100, 176
World War II, 248–249, 257, 259

Yang Hsin Pao, 165

Zambia, 70, 99–100, 103–104, 110, 138, 141–142, 153
Zulu, the, 158, 160, 162
Zululand, 136, 139

73 74 7 6 5 4 3 2 1